Molecular approaches to evolution

Molecular approaches to evolution

Jacques Ninio
National Centre
for Scientific Research, Paris

Translated by
Robert Lang
National Institute for Medical Research,
Mill Hill, London

Princeton University Press
Princeton, New Jersey

First published in the United State of America 1983 by
Princeton University Press,
41 William Street, Princeton, New Jersey 08540

Printed in Great Britain

Library of Congress Cataloging in Publication Data
   Ninio, J.
     Molecular approaches to evolution.
     Translation of: Approches moléculaires de l'évolution.
     Bibliography: p.
     I. Chemical evolution.   I. Title.

ISBN: 978-0-691-64094-5

Translated by Robert Lang from *Approches moléculaires de l'évolution*
© Masson, Paris, 1979
First English edition published by Pitman Books Ltd, 1982

Contents

1 Clearing the ground

We can understand evolution without really knowing what life itself is! This view, held by the great biologist Haldane, was characteristic of a half century of evolutionism which produced models and calculations in the absence of any intimate knowledge of cellular logic. The evolutionist used to try to calculate the outcome of battles for survival just as a General evaluates the chances of victory on a battlefield: will one side score a resounding victory, will the two forces neutralize each other or will they occupy the field alternately? The theory of evolution by natural selection lends itself to all sorts of ramifications. Its principal theme is that since many more individuals of each species are born than can reach reproductive age, those best fitted for the struggle for existence will have more chances of leaving descendants and the characters they carry, passed from generation to generation, will eventually become fixed within the species. If new and even fitter variants appear, they will gradually eliminate the former ones to their own advantage.

I support the doctrine of natural selection in the same way that I recognize the truth of the syllogism that Socrates is a man, all men are mortal, therefore Socrates is mortal – but I go no further than that. A correct idea can become sterile. Renaissance scholars believed that the study of syllogism would enable them to work out the whole of logic. Instead they merely got bogged down in an exploration of Baralipton, Darapti and 254 other outdated forms of syllogism. Of course, evolutionists had to examine closely the fundamental propositions of Wallace and Darwin and assess the possible influences of all imaginable factors: small or large populations, isolated or panmictic, sexed or not. But today we must recognize that this area of evolutionary biology tends to become confined to the study of infinitely increasing variants.

In renewing our conception of living things, molecular biology sheds new light on an evolutionism which until now has been too formal and helps us to redefine the problems. At first molecular biologists were content to verify at their level the validity of concepts developed by the evolutionists 30 years earlier. Duplication of chromosome segments became duplication of DNA segments; comparative anatomy was replaced by comparative molecular anatomy. Fortunately, things did not rest there. After this consolidation phase a new generation of work began which posed the problem of evolution in a completely new manner. In the former view an animal predator with good eyes would be able to capture its prey more easily than one with

1

defective vision and would therefore be selected for. But what made possible the existence of a visual apparatus in animals? Why was this solution realizable and not another? Or, if other solutions were available, what were they? In summary, our aim now is to tackle a problem which not so long ago seemed insoluble: how to evaluate, having taken into account the state of organization of life at any given moment, the various possibilities which are apt to manifest themselves. This is the primary aspect of the new set of problems posed by studying evolution. A secondary aspect is how to decide, once a novelty has appeared, if it will be established or rapidly eliminated. I shall therefore discuss evolution, as far as one can today, in the new perspective as applied to the molecular logic of living things. The genetic code is at the heart of this logic. Above all, we want to know why the code exists and through what tentative stages cellular organization, now so well formed, could have passed. We will tackle step by step all the interesting concepts in molecular biology which revolve around the genetic code.

At first sight the layout and order of the contents will appear disconcerting, but nothing has been left to chance. Those who like to have everything neatly cut and dried will be disappointed. I am not presenting a standard product, aseptic and reassuring, to be offered in a supermarket, but a discourse which sticks to the facts of current research, which refuses to gloss over contradictory aspects of present ideas and which allows, albeit tentatively, social and ideological connotations to show through. Instead of following each statement with ten lines of support, ten against and three of compromise, I concentrate on only one aspect and further on, when I show how the problems can be posed in other ways, I discuss the neglected aspects in other contexts, which lend them their true importance. A major aim of the layout which I have adopted is to produce a text which can be read with profit and interest by readers at very different levels. Technical terms have been kept to a minimum. To avoid tedious exposition of details of molecular biology I have dispersed reminders throughout the text and have placed them in an evolutionary context as soon as they appear.

Initially, of course, the evolutionary approach will be rudimentary: the kind in which molecules from different organisms are compared to deduce ancestral relationships between species. This aspect of molecular evolution has already received wide coverage (it is discussed in Chapter 3). On the other hand, everything else is practically new. No other textbook on evolution deals with comparisons in three dimensions, with acquisitive evolution and with the stability of the genetic code. No other work since the remarkable books of Woese (1967) and Yčas (1969) has discussed the genetic code in depth. After briefly describing constituents (proteins and nucleic acids) and comparing them we shall move on to mechanisms, to the most important cellular processes: information transfer, replication and genetic translation. Having revealed the logic behind these mechanisms, we shall attempt to understand how they could have emerged. Lastly, we shall reach a third stage. The cell is not simply a collection of molecules or a juxtaposition of mechanisms but a coherent entity in which each part is subject to constraints of compatibility with the rest. We shall attempt to clarify this to form concepts which will enable us to understand an evolution which proceeds from one coherent whole to another which is equally coherent.

Free to annoy the purist, I have often preferred to use simple language, close to the spoken word, which is very effective for putting ideas across. The research worker in the laboratory, when he is not writing administrative reports which will be judged by the quality of the typing uses everyday words and images borrowed from daily life. This can hardly be denied by the evolutionists, who learnt selection *from* cattle breeders and took their folklore of family trees from the genealogists. In the *Origin of Species*, Darwin wrote in an accessible style about ideas which were profoundly original for the scientific mentality of his time. Nevertheless, on one point my attitude is different from that of the Master. The progress of ideas is to a large extent a collective endeavour. Although I have contributed to the modern conception of the genetic code, I do not claim any exclusiveness but place my work firmly in the mainstream of modern evolutionary theory.

I thank all those who have helped me in this enterprise, especially the five colleagues and friends who went through the first draft with a fine-toothed comb: Jean-Pierre Dumas, Jerôme Lavergne, Jeanine Rondest, Pierre Roubaud and Michel Volovitch and those who helped me to review the English version: Margaret and Richard Buckingham, Dick d'Ari and Liliane Assairi. I also thank all those who, through the interest which they showed for the subject, encouraged me to set down these reflections on evolution. Above all, my gratitude to Robert Lang: I recognize my child in his translation. Luck or prescience, the French text, written in 1977–1978 needed little modification to cope with the latest advances. The Bibliography was updated in July 1980, then in October 1982, and adapted for an English-speaking audience.

2 **The chemistry of life**

At any given moment thousands of chemical reactions are occuring in the living cell. The food absorbed serves both as fuel, which supplies energy, and as material for constructing new structures or for replacing parts used by the cell. Only plants and photosynthetic bacteria can draw their energy directly from sunlight. The transformations, degradations and syntheses are chemical in nature. Each reaction in the cell is carried out and controlled by a molecule which is designed for this purpose – an enzyme. The enzyme interacts with the substances which have to be chemically combined and accelerates the speed of the reaction much more effectively than any mineral or organic catalyst known to chemists. The general biochemistry by which the cell synthesizes enzymes is common to all terrestrial life forms with only minor variations from the most rudimentary to the most highly evolved creatures.

If reactions were left to occur naturally in water, a molecule A would be able to change into many different molecules: B, C, D, etc. D in its turn would result in L, M, O, P, etc. The enzyme, by accelerating only some of these transformations, orientates the processes along very special routes. Everything done chemically in living creatures is feasible without enzymes. But their presence, by selectively accelerating certain reactions, makes the chemical machine run and keeps it stable in its motion. In chains of reactions in the cell, where the product of each reaction is the substrate of the following one, all intermediates are maintained at workable concentrations. There are neither flat spots nor bursts of power. The speeds of the reactions in a chain and the concentrations of the intermediates are well adjusted to each other. The cellular machinery runs smoothly despite environmental variations.

Another feature of chemical transformations in the cell, which is linked to the first, is that they occur gradually. Reactions which would produce or consume too much energy are split up into a large number of steps. For example, under cellular conditions, the breakdown of glucose to water and carbon dioxide must release about 2.85 MJ/mol (680 kcal/mol). Cellular reactions use energies in the order of about 42 kJ/mol (10 kcal/mol), often less. Reactions which consume energy are coupled to others which produce it; one way of raising a weight is to link it to a counterweight by a pulley or winch and to lower the counterweight. By carefully balancing the pulley it is possible to control better the raising or lowering of the weight. In cellular reactions the counterweight principle is provided by the decomposition of ATP (we shall see later what this molecule is like) to ADP or AMP. The energy released is in the order of 50 kJ/mol (12 kcal/mol). In passing we should mention the existence of a second system based on equilibria of oxido-reduction (the $NADH_2$/NADH system). Glucose, converted into 'energetic small change', provides about 38 molecules of ATP, the universal counterweight. But ATP is not in itself an 'energy capsule'. Every chemical reaction is in principle reversible. If we start with compounds A and B and end up with C and D according to $A + B \rightarrow C + D$, then by mixing C and D at high enough concentrations we must be able to form A and B thus: $C + D \rightarrow A + B$. One can recover energy by breaking down a molecular species – glucose or ATP here – only when the species is in excess of the calculated equilibrium concentration. 'Glucose energy' is not converted into 'ATP energy', but a glucose/CO_2 imbalance is converted into an ATP/ADP imbalance.

The enzyme, like every catalyst, accelerates both forward and back reactions by the same amount so that the concentrations of the reactants tend towards their equilibrium values. If the cell has to convert A into C, then instead of using the natural reaction $A + B \rightarrow C + D$, it can substitute for B a molecule analagous to ATP. The thermodynamic constraints are no longer the same and a better C/A ratio can be obtained. In another phase of the cell cycle it might be advantageous to move the reaction in the opposite direction, from C to A. The old enzyme is put out of service and a new one comes into action which links the energy counterweight in the opposite direction. Regulatory mechanisms allow enzyme activities to be modulated as required. If a product is in excess its formation is usually automatically slowed down, since it inhibits one or more of the enzymes in the reaction sequence which leads to its synthesis. There exists another, more elaborate form of regulation, discovered by Jacob and Monod. The presence of a substrate A produces a signal which switches on the production of an enzyme capable of transforming A. In the absence of A, the synthesis of this enzyme is repressed. Thus, superimposed on the network of chemical reactions in the cell, there is a still more complicated system of regulatory actions, interactions and retroactions which adjust the quantities and activities of each enzyme as functions of the concentrations of cellular compounds.

Enzymes and other cellular macromolecules tend to decompose when in contact with water. The small molecules used by the cell are usually relatively unoxidized. When they are exposed to oxygen in the air they are eventually converted into substances which are of no use for cells. Thus substances which form spontaneously in the terrestrial environment bear little resemblance to biological substances. The great idea of the Soviet scientist Oparin, formulated in 1924, was that the chemical environment which prevailed at the Earth's surface when it was formed could have been very different from that of today. Later Haldane introduced the concept of a primitive terrestrial atmosphere, low in oxygen but rich in reducing gases such as methane and ammonia. Oparin specified major ways in which biologically important molecules could have been formed spontaneously under these primitive conditions.

A quarter of a century later, a young American student, Stanley Miller, constructed a chemical reactor which simulated primitive conditions on Earth. He produced electrical discharges in a flask containing methane and ammonia. Products were formed which were carried away in steam and continued to react in a second flask, which took the place of the primitive ocean. On the seventh day Miller saw that the results were good and stopped the experiment. Deposits had formed, turning the water brown. Miller analysed these after acid degradation and presented his supervisor with results about which he had no need to blush. Several biological substances were detected; notably two amino acids, glycine and alanine. Aspartic and glutamic acids were identified later. Thus Miller in 1953 supplied the first experimental support for the theory of Oparin and Haldane. I had the privilege of meeting Oparin while he was in semi-retirement, at the Bakh Institute, surrounded by venerating collaborators. I found in him an intellectual vigour and a wide-ranging and penetrating vision which are not often met in active scientists. Why had no-one around him in 30 years attempted to perform a Miller's type experiment? He saw this as unnecessary since Miller's experiment was fundamentally equivalent to that of Wohler dating from 1832. And in fact several bridges between the chemistry of life and inorganic chemistry had been established over the previous century. This in no way detracts from Miller's achievement, since his experiment was above all a beginning on which a new branch of science, prebiotic chemistry, was going to be built.

Having explored possible ways of synthesizing amino acids, prebiotic chemistry has succeded in proposing plausible mechanisms for the synthesis of sugars, nucleotide bases, phosphorylated compounds and coenzymes – all the small molecules which play important roles in living cells. These syntheses apparently include the preliminary formation of some highly reactive compounds such as formaldehyde (HCHO) and hydrogen cyanide (HCN). Astronomers, with their radiotelescopes, can detect these compounds in space. A notable result of the last 10 years is that a large number of reactive radicals, whose importance has been emphasized in prebiotic chemistry, have been found in interstellar space in surprisingly high concentrations.

Can we find traces of synthetic pathways of prebiotic chemistry in present-day metabolism? Let us take as an example a metabolic pathway leading to the synthesis of an amino acid X:

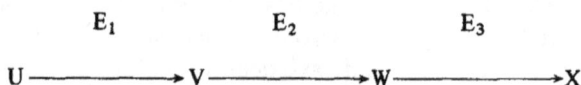

$$U \xrightarrow{\quad E_1 \quad} V \xrightarrow{\quad E_2 \quad} W \xrightarrow{\quad E_3 \quad} X$$

Suppose that the enzyme E_1 catalyses the reaction $U \rightarrow V$, that E_2 catalyses $V \rightarrow W$ and E_3 catalyses $W \rightarrow X$. According to an ingenious argument of

Horowitz's it is reasonable that these three enzymes might have arisen in the course of evolution in the reverse order from that which they have in the scheme. Imagine some very primitive organisms which used X to make their proteins. In those distant times, X was abundant, produced by natural reactions at the Earth's surface, so the cell had no need to synthesize it. Little by little, the reserves of X became exhausted, partly because cells used it up and partly because it decomposed spontaneously according to the sequence: $X \rightarrow W \rightarrow V \rightarrow U$. Then the easiest way of getting X was to synthesize it from its first decomposition product W, hence enzyme E_3 which catalyses $W \rightarrow X$. When W was exhausted in its turn, the appearance of an enzyme E_2 which reformed W from its decomposition product V, together with E_3 which was already present, allowed the lack of X to be made good. The idea we may glean from this hypothesis is that present-day metabolic pathways go in the opposite direction to primitive ones. Other authors, such as René Buvet, see present-day metabolism as a transposition of the great pathways of primitive chemistry. Since the cell often breaks down biological compounds produced by another organism to synthesize closely similar compounds, both points of view have something in their favour. But in the bioenergetic view of the beginning it appears that metabolic pathways obey a third logic: instead of direct and abrupt synthetic routes the cell would prefer gently sloping, winding paths which would allow energetic differences to be split up.

Enzymes are synthesized in the cell by arranging simpler chemical units, amino acids, in a certain order and linking them end to end. Twenty amino acids are used in the synthesis of enzymes and other proteins. Each protein consists of an exact bonding of 50, 100, 1000 or more amino acids, which follow each other in a determined order. Thus human lysozyme, which makes tears effective as a bactericide, starts with the amino acid series: lysine, valine, phenylalanine, glutamic acid, arginine . . . and this sequence is defined until the end. Naturally, other proteins have different amino acid sequences. The chemical procedure by which the cell synthesizes its proteins is extraordinarily precise, as will be shown later. Changing a single amino acid in a protein is sometimes sufficient to change its properties. People with sickle-cell anaemia have an 'abnormal' haemoglobin with a valine instead of a glutamic acid at position 6 in a chain of 146 amino acids. The name enzyme is reserved for proteins with catalytic activity. Proteins also have other roles. Regulatory proteins switch enzyme activity on or off. Proteins are one of the basic materials of all biological tissues: muscle, cartilage, hair etc. In membranes they are in charge of the cells' frontier posts through which products are pumped into the surroundings or excreted. The antibodies (Chapter 14), which protect us against biological aggression, are proteins.

The amino acids, except proline, have the general formula:

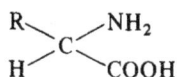

$$\begin{array}{c} R \diagdown \quad \diagup NH_2 \\ C \\ H \diagup \quad \diagdown COOH \end{array}$$

where R is the variable portion or side chain of the amino acid. Two amino acids can be linked, the COOH group of one reacting with the NH_2 group of the other, with the loss of a water molecule. Chain formation by amino acids in proteins follows the principle:

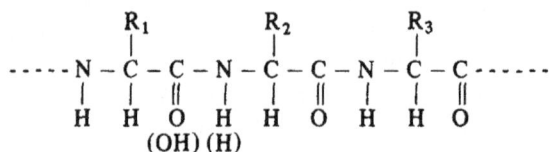

$$
\cdots\cdots N - \overset{\overset{\textstyle R_1}{\displaystyle |}}{\underset{\displaystyle |}{C}} - \overset{}{\underset{\displaystyle \underset{(OH)}{O}}{C}} - \overset{}{\underset{\displaystyle \underset{(H)}{H}}{N}} - \overset{\overset{\textstyle R_2}{\displaystyle |}}{\underset{\displaystyle |}{C}} - \overset{}{\underset{\displaystyle O}{C}} - \overset{}{\underset{\displaystyle H}{N}} - \overset{\overset{\textstyle R_3}{\displaystyle |}}{\underset{\displaystyle |}{C}} - \overset{}{\underset{\displaystyle O}{C}} \cdots\cdots
$$

I have indicated in brackets the atoms eliminated when the link, called a peptide bond, is formed between amino acids. A protein can be represented as a very regular peptide skeleton which bears the amino-acid side groups. Each amino acid has individual features. Cysteine has a thiol group ($-SH$) which allows it to anchor an enzyme to a molecule necessary for its activity and to bridge the different parts of the protein chain. Five amino acids are cyclic; their side chains form rings which restrict their flexibility. These seem to play an important role in interactions between proteins and nucleic acids, which will be discussed later on. Two amino acids, aspartic and glutamic acids, possess a second acid group (COOH). Three have a second basic group (NH_2). An important property of amino acids is their affinity for water. Some of them, the hydrophilic ones, have a tendency to surround themselves with water molecules and dissolve easily in it. Others, the hydrophobic ones, flee from contact with water. This property dictates, to a large extent, the way in which a protein chain is folded in an aqueous solution. The hydrophobic amino acids which flee from water are grouped in the interior while the hydrophilic amino acids face the outside and are covered with water. The size of amino acids varies from $7.8\ nm^3$ for the smallest (glycine), whose side chain consists of a single hydrogen atom, to $249\ nm^3$ for the bulkiest (tryptophan), which contains a heterocyclic group.

Amino acids are composed of only five elements: carbon, nitrogen, oxygen, hydrogen and sulphur, In the structural core common to all amino acids, four atoms or groups of atoms are attached to the central or α carbon. But the groups are not arranged in a plane around the α carbon. They are at the points of a tetrahedron with the α carbon at its centre (Fig. 1). There are only two distinguishable ways of arranging four different objects at the points of a tetrahedron. Of the two symmetrical arrangements shown, only the L-form is found in the twenty amino acids in proteins. The amino acids formed in Miller's experiments contained equal proportions of the L- and D-forms. One of these forms may be favoured chemically only if one starts with compounds in which one asymmetric form predominates; or one has to use extremely complex radiation treatment. The exclusive presence of L-asymmetry in proteins is a puzzling trait of biochemistry. It has long been

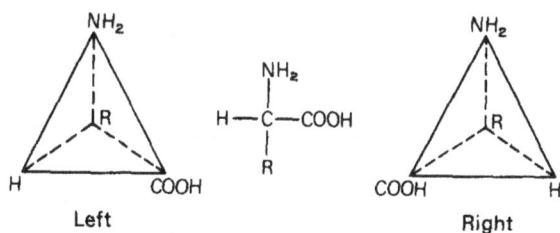

Figure 1 Asymmetry. The structure common to all amino acids (except proline) is shown in the centre. Attached to the central carbon, the α carbon, are four groups: a hydrogen, an acid group (COOH), a basic group (NH₂) and a group R which varies from one amino acid to another. These four groups are not situated in the same plane as the central carbon. They are represented more correctly as being at the points of a tetrahedron at the centre of which is the α carbon. There are two radically different ways of arranging the four groups about the central carbon which are energetically equivalent. Only the L-form is used to make proteins.

considered, wrongly I believe (see p. 50), that this constitutes one of the most fundamental enigmas posed by the origins of life.

In fact some D-amino acids are found in the cell, notably isomers of tyrosine and of leucine which, together with other amino acids, form small circular chains called cyclic peptides. Some antibiotics, such as gramicidin and tyrocidin, are cyclic peptides. Their method of synthesis is completely different from that of proteins, described later. A cyclic peptide is made complete on one enzyme, which achieves the incredibly complex task of choosing the first amino acid then the second, linking them, choosing the third and linking it to the second and so on. Mid-way between small peptides and proteins we find chains of about 20–50 amino acids, such as insulin, which should be classified with proteins because of their method of synthesis.

When amino acids are polymerized in prebiotic chemical experiments they are linked randomly but nevertheless with some types of preferential succession. Steinman finds that these preferred sequences are also those most often found in proteins. This result is not established with certainty and several explanations can be found for it. For example, if certain amino acid sequences were unstable they would have little chance of accumulating in prebiotic environment and they would not be used to make proteins.

Living creatures reproduce: make copies of themselves. Thirty years ago it was possible to believe that this property resided in proteins. An enzyme would interact with the cell membrane, leaving its imprint which would serve as a mould for the synthesis of a new enzyme identical to the first. The idea of molecular reproduction using a mould, initiated by Pauling, turned out to be wrong for enzymes but close to the truth for another class of cellular macromolecules: the deoxyribonucleic acids (DNA), essential constituents of chromosomes. The cell copies only its DNA, which contains instructions in code form for making all of its enzymes. DNA is the molecule of heredity, transmitted from generation to generation. The ribonucleic acids (RNA), chemically closely related to DNA, are used in the making of proteins from the information in DNA, and in certain viruses take the role of hereditary material generally allotted to DNA.

DNA molecules, like proteins, are linear polymers. A DNA chain is made by joining end to end chemical units which show less diversity than amino acids – the four nucleotides: adenine, thymine, cytosine and guanine, which are designated by the letters A, T, C, G. A DNA chain can contain thousands of these units in a determined order: AATCGATTGCC, etc. Nucleotides are

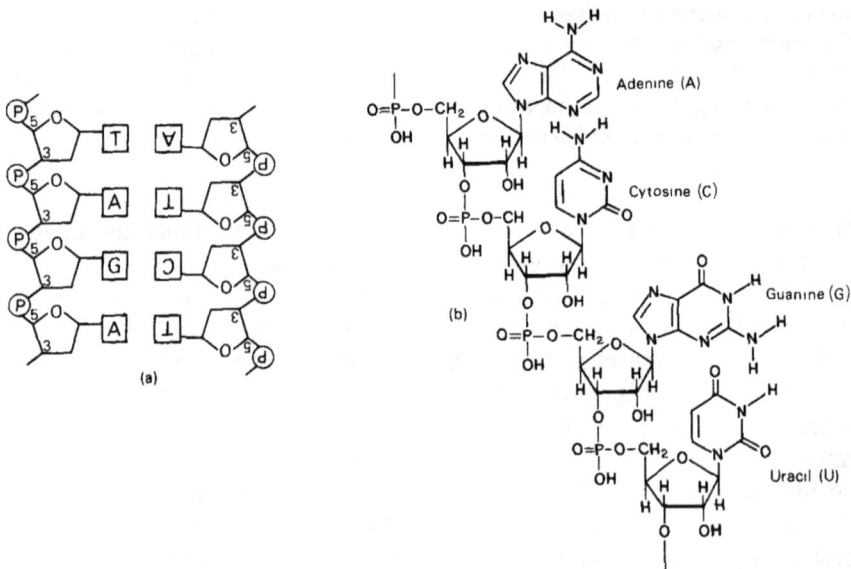

Figure 2 Structure of nucleic acids. (a) Representation of a double chain of DNA. Bases (A, T, C and G) are shown as squares, sugars as pentagons and phosphates as circles. The chemical structure of RNA is shown in more detail in (b). Diagrams (c) and (d) give an idea of the shape of the molecules. In (c) the sugar–phosphate–sugar–phosphate chains are represented by ribbons and the bases by rods. In (d) the picture is more detailed and representative: atomic volumes are shown and the molecule appears as a compact, stratified structure. The stacks are pairs of opposing bases on the two strands. The outer necklaces correspond to the sugar–phosphate chains. The space between the paired molecules on the axis side is called the 'wide groove'. The space on the other side is the 'narrow groove'. In (d) a portion of the 'narrow groove' is shown in the upper part and a portion of the 'wide groove' is below. In (e) is shown in detail how the opposing bases on the two strands match up. The + and − signs indicate the head and tail sides of the base according to an arbitrary convention. Notice the similarity of the external geometries: the two sugars of the G.C. pair are in the same relative positions as those of the A.T. pair. The G.C. pair is shown again more schematically in (f), together with other pairs having their sugars in different relative positions.

much more complex than amino acids and much harder to form in prebiotic chemical reactions. They consist of three parts:

(PHOSPHATE) – (SUGAR) – (BASE)

The base is the variable portion which gives the nucleotide its name (A, T, G or C). The sugar (deoxyribose or ribose) is what distinguishes DNA from RNA. The phosphate is the reactive component by which the monomers are linked to form polymers (*see* Fig. 2). In the cell, free nucleotides may carry one, two or three phosphates. If they have one, as above, they are given the

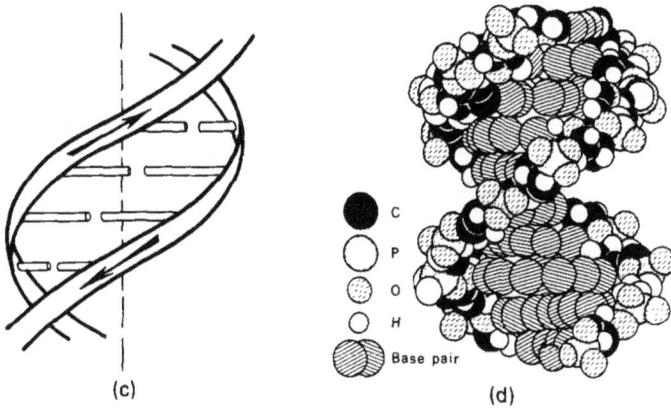

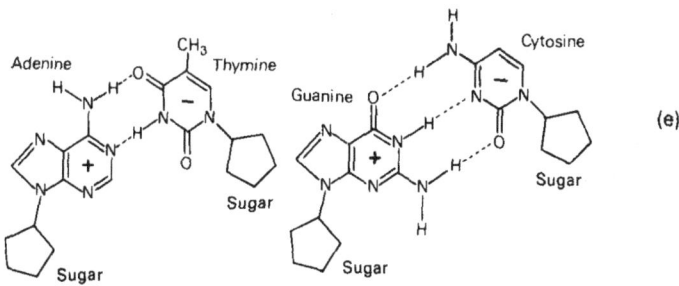

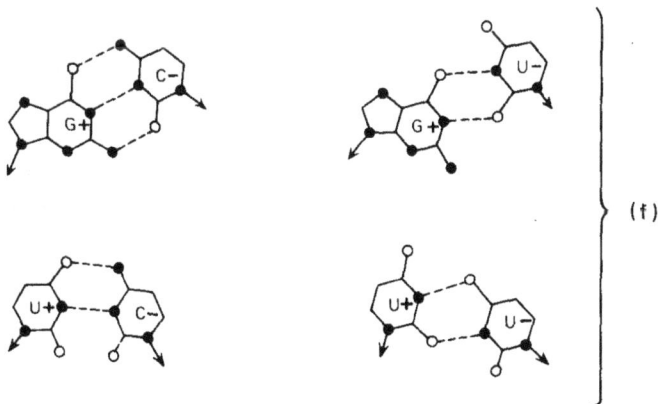

initials AMP, GMP, CMP and TMP (MP for monophosphate). When they have two they are abbreviated as ADP, GDP, CDP and TDP; and those with three phosphates are called ATP, GTP, CTP and TTP. These are the correct forms when the sugar is ribose. For deoxyribose a d is placed in front: dATP, dCTP, etc. ATP is none other than the key molecule of cellular energetics, mentioned at the beginning of the chapter. In RNA, thymine is replaced by uracil (U), a closely related structure.

Nucleic acids are synthesized by an enzyme – DNA or RNA polymerase – which takes a pre-existing DNA or RNA molecule as its model. Let, for instance, . . . AATCGGCATT . . . be the model chain or template. The polymerase makes a sequence said to be complementary to the first, according to the following rules of correspondence: a T on the newly formed chain corresponds to an A on the model (template) and vice versa; a G on the new chain corresponds to a C on the template and vice versa. A and T constitute a complementary base pair as do C and G. Later on, the newly-formed chain is used as a model. Performing these complementary operations twice (A → T then T → A) results in an exact copy of the original sequence. The polymerase moves along the template, which it copies, and has to 'select' the nucleotide corresponding to each base of the template. In fact the enzyme does not select, its role is passive. The nucleotides have affinities for one another. Thus G has a strong affinity for itself and for C and a weaker affinity for T. C has a particularly strong affinity for G. T can associate with T, C, G but especially with A; while A, according to circumstances, pairs with T, C or with itself. All these associations, relatively unstable in water, become much more stable in environments from which water is excluded. The major role of the polymerase is apparently to provide a hydrophobic environment in which a nucleotide of the template, T for example, transitorily attracts a free nucleotide, A or G (to be more precise it is the triphosphate forms which participate in this process).

When the pair is formed there is, on the one hand, the elongating chain whose end occupies a well-defined position in space and, on the other hand, the associated nucleotide whose active part (by which the link can be made with the chain) is at a variable distance from the other. The link will be made only if the groups are at the right distance. The four complementary associations A.T, T.A, C.G and G.C have the remarkable property of bringing the active group into the same position in space. They are geometrically interchangeable; their similarity is shown in Fig. 2(e). The polymerase, without recognizing the nucleotides individually takes hold of the active group when it is at the standard position for complementary pairing and joins it to the elongating chain. Since each nucleotide of the template has an affinity for its opposite number in the complementary chain, the two strands tend to remain associated, forming a double chain:

```
. . . . A A T C G G C A T T . . . . .
. . . . T T A G C C C T A A . . . . .
```

DNA is nearly always in the form of a double chain. During replication, the two chains separate along part of their length and DNA polymerases synthesize chains complementary to both strands at the same time (*see* Chapter 6). Each new chain remains attached to the one which served as its template. The result of this is that the initial double chain is doubled. RNA is also synthesized by complementary replication, generally from a DNA template. But it does not remain attached to its template. Instead it is picked up by other cell components. So RNA exists more often as a single strand but the strand is folded up with the nucleotides pairing as well as they can to form associated regions of varying lengths. Often regions of the RNA chain which self-associate are complementary or almost complementary sequences. In addition to complementary associations, dominated chiefly by the pairs G.C, A.U and to a lesser extent G.U, there are associations of different types; for example, between two regions rich in G residues.

Nucleotide bases are flat molecules with head and tail sides which are stacked on one another in nucleic-acid chains. Let us now consider DNA not as an association of two chains (template and copy), but as a polymer whose elementary unit is the complementary base pair. Owing to the geometric similarity of the complementary pairs, DNA can be thought of as a repeated single geometric unit. Now, when a polymer is formed so that the spatial relation between the first and second elements is exactly the same as that between the second and third and so on, a helical arrangement necessarily results. A protein made of only one amino acid is a helix. DNA, which is a succession of geometrically equivalent base pairs, is a (double) helix.

The DNA double helix described by Watson and Crick in 1953 captured the imagination of biologists. It soon became the symbol of molecular biology, the banner of a whole generation of biologists. The double helix embodies both the dawn of a new era in the life sciences and, less obviously, faith in molecular structure as the ultimate explanation of life. Biology used to be above all a science of observation; after several decades of detours through a rather too abstract biochemistry, DNA brought us back to the visible.

How is the relationship between DNA, the molecule of heredity, and proteins established? Let us say a few words about this before taking the question up again in Chapter 9. The amino acid sequences of proteins are inscribed in DNA in code form. Each protein is coded for by a region of DNA which corresponds to it: its gene. The succession of nucleotides in the gene specifies the succession of amino acids in the protein coded for by the gene according to a set of particularly simple rules, although the cellular apparatus which does the decoding is very complex. During decoding a gene sequence (for example, GGAGAACGCCAC) is taken in consecutive groups of three bases (GGA, GAA, CGC, CAC, etc.) and each of these triplets, called codons, designates a particular amino acid. The meaning of each codon is the same for all living creatures with few known exceptions (*see* Table I). GGA designates glycine, GAA glutamic acid, CGC arginine, etc. In the presence of

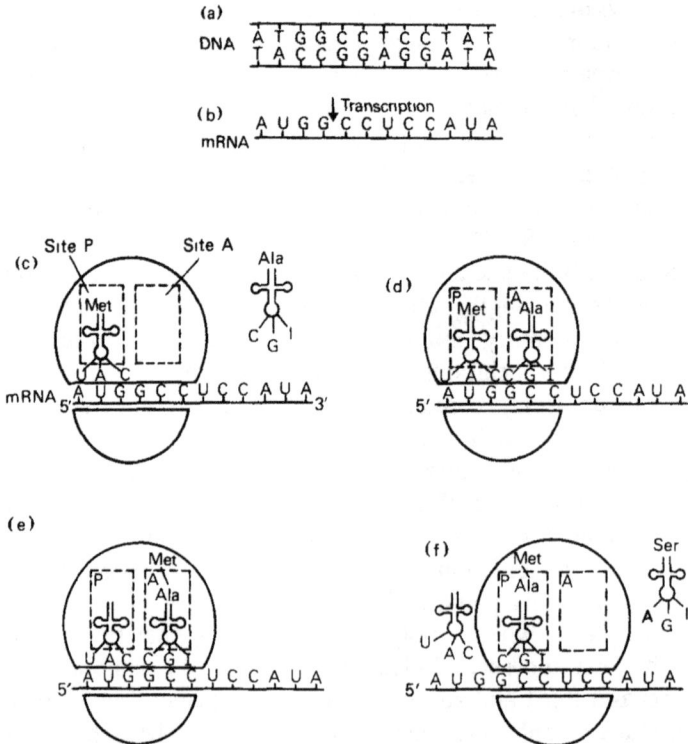

Figure 3 Biosynthesis of proteins. Messenger RNA (b), transcribed from a double chain of DNA (a), is picked up by a ribosome (c). The exact positioning of the ribosome at the site at which reading starts (initiation) requires a complex sequence of events which is not given in detail here. Recognition between the codon which is to be read and the tRNA amino-acid carrier occurs at the A site (d). While the tRNA is kept at site A, the elongating peptide chain is transferred from the tRNA which had read the preceding codon to the tRNA at site A (e). The previous tRNA leaves the ribosome, and the one at site A then passes with the chain which it is carrying to site P (f). A new codon appears at the A site and the process is repeated

the nucleotide sequence given as an example here, the translation apparatus synthesizes the peptide sequence glycine–glutamic acid–arginine, etc. Amino acids can be represented by more than one codon and the complete correspondence between codons and amino acids is given in Table 1. Signals for stopping translation exist. There are three termination codons, also called nonsense codons. Genes nearly always start with a codon ATG or GTG, read as methionine, then specify, codon by codon, the amino acids in a protein and end with one or two termination codons. Generally they are flanked by control sequences which bind regulatory proteins and indicate where synthesis of RNA chains should start or stop. Before being translated the gene is copied in the form of an RNA chain called messenger RNA.

Table 1 The Genetic Code

Second base

		U	C	A	G	
First base	**U**	UUU ⎱ Phenyl- UUC ⎰ alanine UUA ⎱ Leucine UUG ⎰	UCU ⎫ UCC ⎪ Serine UCA ⎬ UCG ⎭	UAU ⎱ Tyrosine UAC ⎰ UAA ⎱ Termina- UAG ⎰ tion	UGU ⎱ Cysteine UGC ⎰ UGA Termination UGG Tryptophan	U C A G
	C	CUU ⎫ CUC ⎪ Leucine CUA ⎬ CUG ⎭	CCU ⎫ CCC ⎪ Proline CCA ⎬ CCG ⎭	CAU ⎱ Histidine CAC ⎰ CAA ⎱ Glutamine CAG ⎰	CGU ⎫ CGC ⎪ Arginine CGA ⎬ CGG ⎭	U C A G
	A	AUU ⎱ AUC ⎬ Isoleucine AUA ⎰ AUG Methionine	ACU ⎫ ACC ⎪ Threonine ACA ⎬ ACG ⎭	AAU ⎱ Asparagine AAC ⎰ AAA ⎱ Lysine AAG ⎰	AGU ⎱ Serine AGC ⎰ AGA ⎱ Arginine AGG ⎰	U C A G
	G	GUU ⎫ GUC ⎪ Valine GUA ⎬ GUG ⎭	GCU ⎫ GCC ⎪ Alanine GCA ⎬ GCG ⎭	GAU ⎱ Aspartic GAC ⎰ acid GAA ⎱ Glutamic GAG ⎰ acid	GGU ⎫ GGC ⎪ Glycine GGA ⎬ GGG ⎭	U C A G

Third base

* For a long time, the universality of the code was an article of faith. Some exceptions are now coming to light. UGA is read as tryptophan in human and yeast mitochondria. In the latter CUN codons are translated as threonine. In some mitochondria, AUA codes for methionine. In some organisms, there seems to be a mechanism for the specific incorporation of seleno-cysteine, a cysteine in which selenium has replaced sulphur.

Often a single messenger RNA molecule carries copies of several genes which follow each other. A group of genes transcribed into RNA in a co-ordinated manner forms an 'operon'. Part of cellular regulation is carried out by regulatory proteins (repressors) which act not at the individual gene level but on the whole operon. Translation of messenger RNA is performed on a large macromolecular complex, the ribosome. One of the sites of the ribosome

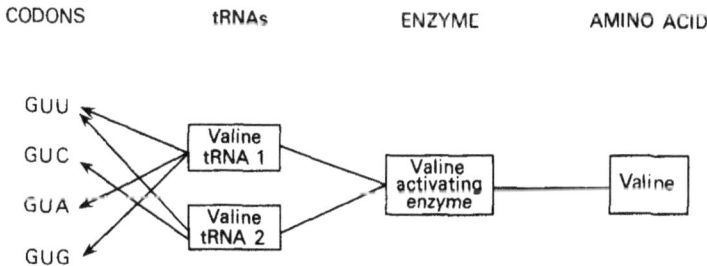

Figure 4 *Unique or degenerate correspondence.* For each amino acid there is only one activating enzyme which can link it to one or more tRNAs specific for the amino acid. A tRNA can read one or several of the codons for the amino acid which it transfers.

functions as a playback head to which the codons on the RNA are exposed in turn. The amino acid which corresponds to the codon presents itself, transported by a molecule of a special class called transfer RNA. Each transfer RNA (tRNA) transports a single sort of amino acid and is also capable of recognizing at least one of the codons of the transported amino acid. When a codon is displayed at this site, one of the tRNAs capable of recognizing it binds to it and the amino acid which it carries is transferred to the growing protein. Further details are shown in Figs. 3 and 4. The link between a tRNA and its amino acid is made by a class of proteins, the amino-acid activating enzymes. There are twenty activating enzymes, one per amino acid. Each activating enzyme attaches its amino acid to the tRNA or tRNAs which corresponds to it.

There is an additional complication to this scheme in eukaryotes – the organisms whose cells have a nucleus. The gene coding for a protein can be made of a number of non-contiguous portions. We have in DNA a succession of coding sequences (exons) and non-coding ones (introns). After transcription, the corresponding messenger RNA is processed so that the introns are spliced out, and the coding parts of the gene form a continuous sequence, suitable for being translated by the ribosome.

Regulations and energetic couplings which make the cellular machine work in a coherent manner, the DNA double helix and replication based on complementary associations, genetic coding between nucleic acids and proteins: such are the great secrets of life at the molecular level. *Se non è vero, è ben trovato!**

---

* If it is not true, it is a clever invention.

3 The discreet charm of the sequences

Atlas of Protein Sequence and Structure: this is the name of the molecular evolutionist's Bible of the 1960s, the great Book of Life in which all known protein sequences are to be found. To save space, each amino acid is represented by a single letter: V for valine, M for methionine, etc. Phrases such as KLPP(M,N,O)PR, etc. cover whole pages, the brackets and commas indicating uncertainties. One shudders to think of the next edition of this work, which could contain 500 pages of this prose, representing thousands of years of human labour invested in the determination of structures. But 10 000 such volumes would be necessary to include the DNA sequence of a single man. By comparing all known sequences, the two compilers of this album, Margaret Dayhoff and Richard Eck, hope to reveal evolution's grand design, to deduce the sequences of primitive proteins which have now disappeared. Those who have put so much care into this remarkable working tool may easily be forgiven their oversimplified approach to the genetic code and its origins.

Stretches of DNA totalling several kilonucleotides are being sequenced every year, not always too accurately. The sequences are stored in data banks and travel as magnetic tapes, or as messages from computer to computer. Programs to handle and analyse the sequences are becoming an object of trade. If the trend is pursued, more and more results will be obtained using commercial computer programs whose hidden assumptions are unknown to the user and often unclear to the conceptor.

I will discuss what is known about protein sequences, using the haemoglobins as an example. Man possesses six different chains of haemoglobin which are given the Greek letters α, β, γ, δ, ε, τ. The last two, which function during fetal life, are hardly studied. Four chains associate to form the active haemoglobin molecule and the most common combinations have the formulae $\alpha_2\beta_2$, $\alpha_2\gamma_2$ and $\alpha_2\delta_2$. The formula $(\alpha\beta)_2$ would be more appropriate, since it suggests correctly that the enzyme is formed from two spatially equivalent halves $(\alpha\beta)$. The sequences of the four chains α, β, γ, δ are given in Table 2. They have roughly the same lengths: 141 amino acids in the α chain, 146 in the other three. The sequences show strong analogies. For each pair of sequences, more than one-third of the corresponding sites are occupied by the same amino acid; the coincidence is too great to be accidental. It seems entirely reasonable that the four chains evolved by mutation from a common

Table 2   Amino acid sequences of haemoglobins and human myoglobin. The sequences are set out so that the maximum number of amino acids coincide. This necessitates artificially interrupting the sequences and shifting them along, which is indicated by dashes.

1 α human	V - L S P A D K T N V K A A W G K V G A H A G E Y G A E A L E R M F L S F P T T K T Y F P H F - D L S H
2 α dog	V - L S P A D K T N V K S T W D K I G G H A G D Y G G E A L D R T F Q S F P T T K T Y F P H F - D L S P
3 β human	V H L T P E E K S A V T A L W G K V - - N V D E V G G E A L G R L L V V Y P W T Q R F F E S F G D L S T
4 δ human	V H L T P E E K T A V N A L W G K V - - N V D A V G G E A L G R L L V V Y P W T Q R F F E S F G D L S T
5 β dog	V H L T A E E K S L V S G L W G K V - - N V D E V G G E A L G R L L I V Y P W T Q R F F D S F G D L S T
6 γ human	G H F T E E D K A T I T S L W G K V - - N V E D A G G E T L G R L L V V Y P W T Q R F F D S F G N L S S
7 Human myoglobin	G - L S D G E W Q L V L N V W G K V E A D I P G H G Q E V L I R L F K G H P E T L E K F D K F K H L K S

1 α human	- - - - G S A Q V K G H G K K V A D A L T N A V A H V D D M P N A L S A L S D L H A H K L R V D P
2 α dog	- - - - G S A Q V K A H G K K V A D A L T T A V A H L D D L P G A L S D L H A Y K L R V D P
3 β human	P D A V M G N P K V K A H G K K V L G A F S D G L A H L D N L K G T F A T L S E L H C D K L H V D P
4 δ human	P D A V M G N P K V K A H G K K V L G A F S D G L A H L D N L K G T F S Q L S E L H C D K L H V D P
5 β dog	P D A V M S N A K V K A H G K K V L T S L G D A I K H L D D L K G T F A K L S E L H C D K L H V D P
6 γ human	A S A I M G N P K V K A H G K K V L T S L G D A I K H L D D L K G T F A Q L S E L H C D K L H V D P
7 Human myoglobin	E D E M K A S E D L K K H G A T V L T A L G G I L K K K G H H E A E I K P L A Q S H A T K H K I P V

1 α human	V N F K L L S H C L L V T L A A H L P A E F T P A V H A S L D K F L A S V S T V L T S K Y R - - - - - -
2 α dog	V N F K L L S H C L L V T L A C H H P T E F T P A V H A S L D K F F A A V S T V L T S K Y R - - - - - -
3 β human	E N F R L L G N V L V C V L A H H F G K E F T P P V Q A A Y Q K V V A G V A N A L A H K Y H - - - - - -
4 δ human	E N F R L L G N V L V C V L A R N F G K E F T P Q M Q A A Y Q K V V A G V A N A L A H K Y H - - - - - -
5 β dog	E N F K L L G N V L V C V L A H H F G K E F T P Q V Q A A Y Q K V V A G V A N A L A H K Y H - - - - - -
6 γ human	E N F K L L G N V L V T V L A I H F G K E F T P E V Q A S W Q K M V T G V A S A L S S R Y H - - - - - -
7 Human myoglobin	K Y L E F I S E - C I I Q V L Q S K H P G D F G A D A Q G A M N K A L E L F R K D M A S N Y K E L G F Q G

A = Alanine
C = Cysteine
D = Aspartic acid
E = Glutamic acid
F = Phenylalanine

G = Glycine
H = Histidine
I = Isoleucine
K = Lysine
L = Leucine

M = Methionine
N = Asparagine
P = Proline
Q = Glutamine
R = Arginine

S = Serine
T = Threonine
V = Valine
W = Tryptophan
Y = Tyrosine

ancestor. The family resemblance becomes even more striking when we look at the shapes in space. All the haemoglobin structures which have been established by X-ray studies can be superimposed on each other in space (Fig. 8). The chains are similarly folded, form the same hollows and associate in the same fashion. Another protein, myoglobin, found in muscle, has a three-dimensional structure which coincides with that of an isolated haemoglobin chain. The sequences of the myoglobins and haemoglobins are sufficiently similar for all biologists (after Zuckerkandl and Pauling) to have declared allegiance to the idea of a protein ancestor common to the myoglobins and the haemaglobins. The balance-sheet of differences between the human chains is as follows:

$$
\begin{array}{lcccc}
\alpha & 119 & & & \\
\gamma & 118 & 89 & & \\
\beta & 118 & 84 & 39 & \\
\delta & 117 & 85 & 41 & 10 \\
& \text{myo.} & \alpha & \gamma & \beta
\end{array}
$$

The numbers above were calculated from the sequence alignment presented in Table 2. For each pair of sequences, one counts the number of positions in which they differ either because the amino acids are not the same or because an amino acid faces a space (deletion). Until now, I have spoken of the proteins from one species: man. All haemoglobins and myoglobins in the animal kingdom can be included in the same family as the human haemoglobins. Chimpanzee haemoglobin is identical to man's. Human and horse α haemoglobins are identical in 123 positions and differ in eighteen. These two α chains are closer to each other than a pair of α and β chains from the same organism, man or horse. We can construct a scenario to explain this peculiarity. Long ago there lived an animal which was a distant ancestor of man and the horse. It possessed a haemoglobin whose gene duplicated. One example of the gene gave rise to the α-haemoglobin line and the other to the β-haemoglobin line. Mutations accumulated in the two lines, α and β, so that we now find about ninety differences between the two chains whatever the individual who carries them. Later on, well after the doubling episode, one of the species which possessed the two haemoglobins split into two branches, one leading to man and the other to the horse. At the time of the split the ancestors of man and the horse were still members of the same species: there is only one ancestral α molecule for man and horse and one ancestral β molecule. After the divergence each branch evolved independently and the two lines accumulated mutations. Let us suppose that overall the α and β sequences evolve at the same speed. Then the number of accumulated differences between the human and horse α chains should roughly equal that of the differences between the β chains of the same two species. This is just what is found: eighteen differences in one case, twenty-five in the other. Values for differences between chains systematically show this type of regularity which ties in well with the general scheme of divergent evolution beloved of the Darwinists and whose symbol is the phylogenetic tree (Fig. 5) linking all species. If we look carefully at the table of differences between the human chains, an interesting property emerges. Imagine forty residues of a sequence A which includes 150 amino acids are mutated. A sequence B is obtained and forty of its residues are mutated at random, which gives a sequence C. Overall, we can predict that between sequences A and C there will be somewhat less than eighty differences (since some residues will have mutated twice); let us say seventy differences. The 'distance' relationships between proteins A, B and C would thus be:

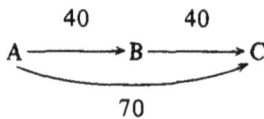

$$A \xrightarrow{\quad 40 \quad} B \xrightarrow{\quad 40 \quad} C$$
$$\underset{70}{A \underrightarrow{\qquad\qquad\qquad} C}$$

Sequence B is half-way between sequence A and sequence C. Now, this kind of situation is almost never encountered in real proteins. There are no 'in-between' sequences. For human myoglobin, all human haemoglobins are equivalent; they differ from it in 118 ± 1 positions. We can thus establish a dichotomy which radically separates myglobin from the four haemoglobins, and there is no protein half-way between the two classes among all those which have been sequenced in the whole animal kingdom. For the α chain, all other haemoglobin chains are equivalent and differ from it by 86 ± 3 residues. No sequence is found intermediate between that of the α chain and the other three. This type of result is not evident *a priori*. The lamprey is a primitive being, man is a highly evolved animal and the carp an intermediate creature. The three species can be placed on successive levels of an evolutionary ladder. This would be anatomically justifiable, but: we cannot order the species in this manner using their α-chain sequences. The commonly accepted interpretation, which I share, is that species split and after each divergence the two species resulting from a common ancestor continue to change. Even if one of them retains morphological traits of the ancestor, its molecules change. The distance relationships are then better described by a phylogenetic tree such as that shown in Fig. 5(a).

Comparisons of sequences of cytochromes *c* – proteins used in electron transport – provide similar lessons. The proteins are very homogeneous in size and the distances between cytochrome sequences obey the law of dichotomy (the 'ultrametric' law) described for the haemoglobins. But the variability is more restricted. On average, according to Dayhoff, 3 per cent of the residues of a cytochrome change in 100 million years as against 12 per cent for a haemoglobin. Histones, proteins which are attached to DNA in

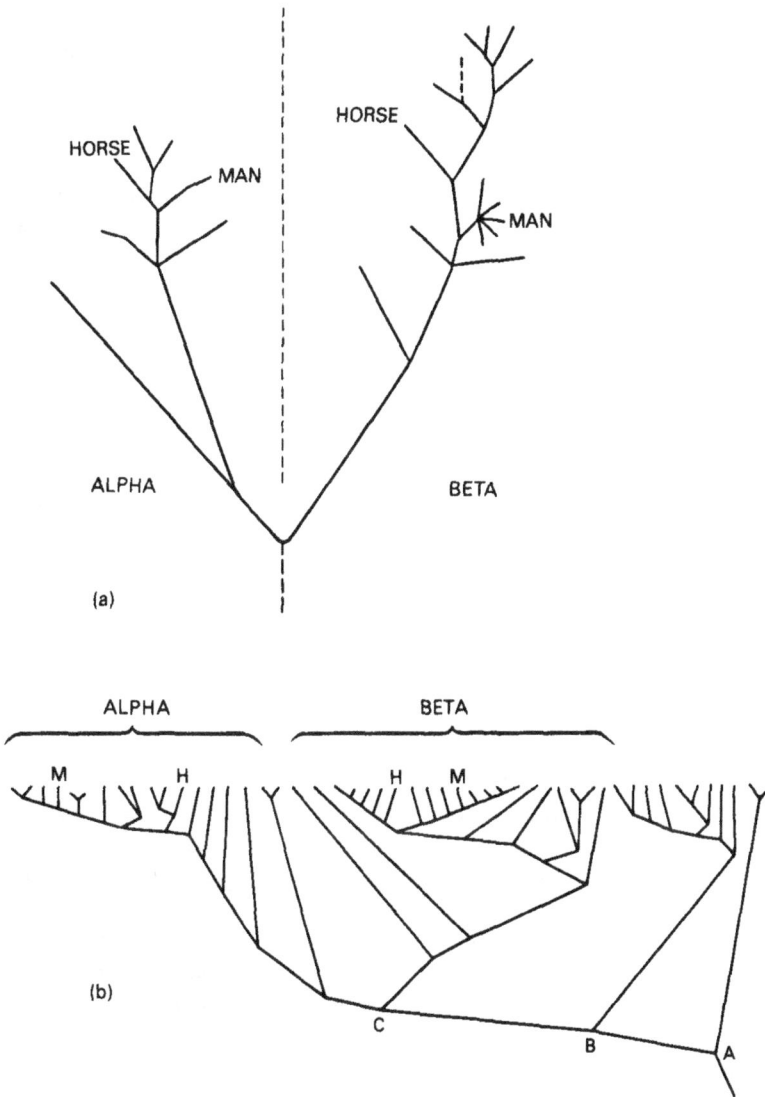

Figure 5 Two phytogenetic trees constructed from haemoglobin chains. To avoid giving too much detail, only the positions of the α and β chains of man and the horse are indicated. The length of each branch is assumed to be proportional to the number of mutations accumulated and fixed along the branch. In the first tree (a), made by Eck and Dayhoff (1969), the β chains seem to have undergone more mutations than the α chains. But the number of β haemoglobin sequences taken into account was twice that of α sequences. In a more recent edition of the Atlas, the numbers of α and β sequences were balanced and the branches have become roughly equal. In the tree of Goodman, Moore and Matsuda (1974) (b) the α chains seem to have mutated more often than the β chains.

chromosomes, change even more slowly: 0.06 per cent change on average in 100 million years. Histone genes, however, mutate faster but their variation is not echoed in histone sequences, owing to the degeneracy of the genetic code. The same amino acid, at homologous positions in two different histones, is generally specified by different codons in the two genes. As no mechanism exists for preferentially mutating codons into synonymous codons, many mutations must be produced which would alter the nature of the amino acid but which are not retained: they are selected against. The same phenomenon is found, though attenuated, in the haemoglobins. We know the sequences of the genes coding for the β chain in man and the rabbit. There are forty-eight nucleotide differences between the genes, which are mostly silent since the proteins differ in only fourteen positions.

By extending the comparison of known sequences in a family of proteins a stage further, we might hope to deduce the structure of the ancestral protein and synthesize it in the laboratory. We would then be in a position, envisaged by Zuckerkandl and Pauling, to study experimentally the properties of proteins belonging to creatures which have been extinct for hundreds of millions of years. And if we established the sequences of the haemoglobin ancestor and the cytochrome ancestor, would we find any similarity between them? It seems that the two families are quite unrelated. Neither the sequences nor the three-dimensional structures allow us to infer the slightest relationship.

When a gene doubles, both copies at first code for the same protein. Later the sequences may diverge, as we have seen for haemoglobin chains. The divergence may go even further so that while one gene continues to code for the usual protein the other mutates to the point where it specifies a protein able to perform a new function. But the two genes may also fuse after the duplication producing one coding for a protein twice as long as the original one. If mechanisms of this kind operate it should be possible to detect internal repeats in sequences of certain proteins. And in fact a good number are known. Human haptoglobin α² has 142 residues. The sequences between residues 13–71 and 72–130 are identical. All large amino-acid activating enzymes show very clear internal repeats. When the doubling is a very old event, the two halves which fused have been able to change in different ways. This makes the internal repeat less easy to detect, as in the example below:

```
... A Y K I - A D S C V S C G A C A S E C P V N A I S Q G ...
... I F V I D A D T C I D C G N C A N V C P V G A P V Q E ...
```

These are portions of the sequence (residues 1–26 and 29–55) of a ferredoxin of *C. pasteurianum* which is often given as an example of an 'archaic' protein. Yčas considers that the above segments were themselves formed by repetition of even shorter sequences. According to him, very many primitive proteins were periodic, reproducing the same short sequence ten, twenty or 100 times.

We know some proteins which are made like this; they form the framework of certain tissues but have no catalytic activity. Thus collagen, a constituent of cartilage, is a protein with a theme (proline–proline–glycine) repeated with variations 300 times.

What are the detailed amino acid substitutions which we observe? Substitutions can be directive or non-directive. When we study variants of human α haemoglobin, notably in anaemics who carry a mutant form, we characterize the mutations with an arrow thus: $Arg_{92} \rightarrow Leu$, which means that the residue at position 92, which is normally arginine, has mutated into leucine. In 1976 about 230 variants of human α and β haemoglobins were known. Almost always, the variant is simple and differs from the normal chain by only a single residue.

This fact is far from trivial. Let us go back 100 000 years and imagine that we have a situation analogous to the present one with one very predominant form of haemoglobin and many variants which each differ from it by a single amino acid substitution. Let us also suppose that, as now, mutations occur anywhere along the chain. If two variants are taken, one differing from the major form in position 35, for example, and the other in position 122, and compared with each other they are shown to differ not in one but in two positions. At the moment when one of the variants has just imposed itself it must differ on average by two positions from the other variants of 100 000 years ago.

As double variants of the most widespread form are never observed, we must assume that all the others have been eliminated. The former companions were eliminated, for being less fit, or simply for being too few: the 'life expectancy' of a minor variant in a population is all the lower for being less widespread since the slightest accident might make it disappear completely.

A list of seventy types of substitution has been gathered from the 230 variants of human haemoglobin. The variant can nearly always be explained by a single base change in the codon. The table of substitutions is quite symmetrical. Major exceptions to this include glycine and arginine. There are thirteen substitutions in the direction glycine→ aspartic acid, eight in the direction glycine→ arginine against three and zero, respectively, in the opposite directions. There are six substitutions of arginine for other amino acids but thirty-three substitutions in all for arginine. This asymmetry for arginine can be partly explained. Arginine codons of the type CGX are rarely used. So there are few substitutions which start off from these codons. But in the short term nothing prevents the codons of other amino acids from mutating into the CGX codons of arginine. The observed amino acid substitutions are often drastic, for these are the substitutions which biochemists detect most easily.

If we compare haemoglobins from different species – carp, horse and man – we have no criteria for judging the direction of amino acid substitutions. We are obliged first to infer the ancestral sequences; substitutions for these, giving contemporary sequences, will then follow with time. The table of directive substitutions appears reasonably symmetrical with a very few exceptions which could be significant, notably a tendency to an increase in arginine residues (Gautier). In practice, we are reduced to speaking of amino acid substitutions between two sequences without attributing direction to them. A good third of the substitutions observed necessitate changing more than one base in the codons. Moreover, the changes in the physico-chemical properties of the amino acids are much less severe than those seen in abnormal haemoglobins. An 'index of similarity' between pairs of amino acids developed in Lyon by Grantham takes account of three properties: chemical composition, charge and volume. The amino acids which are most similar, according to these criteria, are also those which are most often substituted for each other.

Transfer RNAs, like the haemoglobins, form a family in two ways. Each cell possesses about forty types of tRNA, which show similarities. The tRNAs of one species have sequences very close to those of tRNAs in related species. They are built according to the same general model: the cloverleaf structure, shown in Fig. 6, p. 26, which includes several regions with complementary pairing. The precise spatial arrangements are known for a few tRNAs (see Fig. 16e, p. 88) and show considerable similarity in three-dimensions.

If we attempted to reconstruct the scenario of tRNA evolution from sequence comparisons, as with the haemoglobins, we would be disappointed. The pack has been shuffled too much, and we can guess why. In the cell, the tRNAs form a family which is both very homogeneous and quite diversified. In many respects they constitute the interchangeable pieces of a game. They each take in charge an amino acid, associate with an 'elongation factor' – the same for all tRNAs – bind to the same ribosomes at the same site and transfer their amino acids to the peptide chain in the same way. Then, each tRNA has to be distinguished from all other tRNAs which do not bind the same amino acid. Otherwise it would be charged incorrectly by the wrong amino-acid activating enzyme. The tRNA is thus subject to both homogenizing and diversifying constraints. From the foregoing, it follows that a mutation in a tRNA sequence is rarely an isolated event. Each change must be followed by adjustments so that the tRNA continues to obey as well as possible all the constraints.

A tRNA which is specific for a given amino acid can, after point mutations, become specific for another amino acid and this may happen at any time in evolution. It is perfectly conceivable that in *E. coli* a glycine tRNA gene duplicates and one of the genes mutates to code for a valine tRNA and that another valine tRNA falls into disuse, disappearing in favour of the new one. In another species the homologous tRNA presently in use might very well have originated from a threonine tRNA.

The range of sequence variation in tRNA is much more limited than for proteins of a similar length. At each position there is a choice of only four nucleotides instead of one between twenty amino acids. Thus, we suppose that the possibilities of changes in tRNA sequences have already been well explored and that as a result tRNA evolution stagnated.

A way out of this impasse may have consisted in chemically modifying tRNA molecules after their transcription. Methyl groups are added on at several points in the sequence. According to the particular case, uracil is replaced by dihydrouridine, 2-thiouracil, thymine or pseudouracil, etc. In fact the best way of making interspecies comparisons of tRNAs intelligible is to base them on details of modified bases and not on sequence homologies. In this way comparative molecular anatomy – or as Zuckerkandl simply calls it, chemical paleogenetics – resembles the comparative anatomy of the zoologists and botanists. The study of modified bases in tRNAs clearly reveals the split between primitive species (prokaryotes, whose cells have no nucleus) and species with nucleated cells. Chemical modifications of tRNAs are well correlated with the position of the organism in the hierarchy of living forms. Mammalian tRNAs are highly modified, those of bacteria are less so and the tRNAs of mycoplasmids (types of miniature bacteria) are less modified still. Chloroplasts, which are a kind of 'cell within the cell' in eukaryotes, have their own tRNAs which resemble those of bacteria much more than the tRNAs of other compartments in the same host cell (*see* Chapter 15). The tRNAs of mitochondria, other cells within the cell, seem to form a completely separate class no more closely related to bacterial tRNAs than to eukaryotic tRNAs.

Another class of RNA, whose sequences can more happily be compared, is that of the ribosomal RNAs. The ribosome nearly always contains three kinds of RNA (but only two in mitochondria) which are named according to their lengths: 5S for the shortest (120 nucleotides in *E. coli*), 16S (12 times as long) and 23S for the longest (2900 nucleotides or more). Ribosomal RNAs, like the tRNAs, are present in all living creatures; their sequences vary little so we can compare distantly related species. We know about fifty 5S RNA sequences for bacteria, algae, fungi, animals and higher plants. These comparisons lead to the joyful conclusion that man appears to be closer to the hen than to the toad. Complete sequences of 16S RNA are scarce. But we can compare fragments of sequences from various species. Woese's team, who are carrying out this work, have succeeded in producing a classification encompassing all living things. In this, the chloroplasts again appear to be close to the bacteria. A class of bacteria which seem very archaic because of their metabolism – the methanogens – have a ribosomal RNA as far removed from bacterial ribosomal RNA as it is from eukaryotic ribosomal RNA.

Woese boldly proposed that we were in the presence of a third form of life – and that the former dichotomy in prokaryotes and eukaryotes should be replaced by a three-way classification, comprising three kingdoms: the eukaryotes, the standard bacteria, and the archaebacteria. His classification holds good. The archaebacteria now include most of the bacteria that live in extreme environments: the thermophiles which grow in hot springs at 80 or

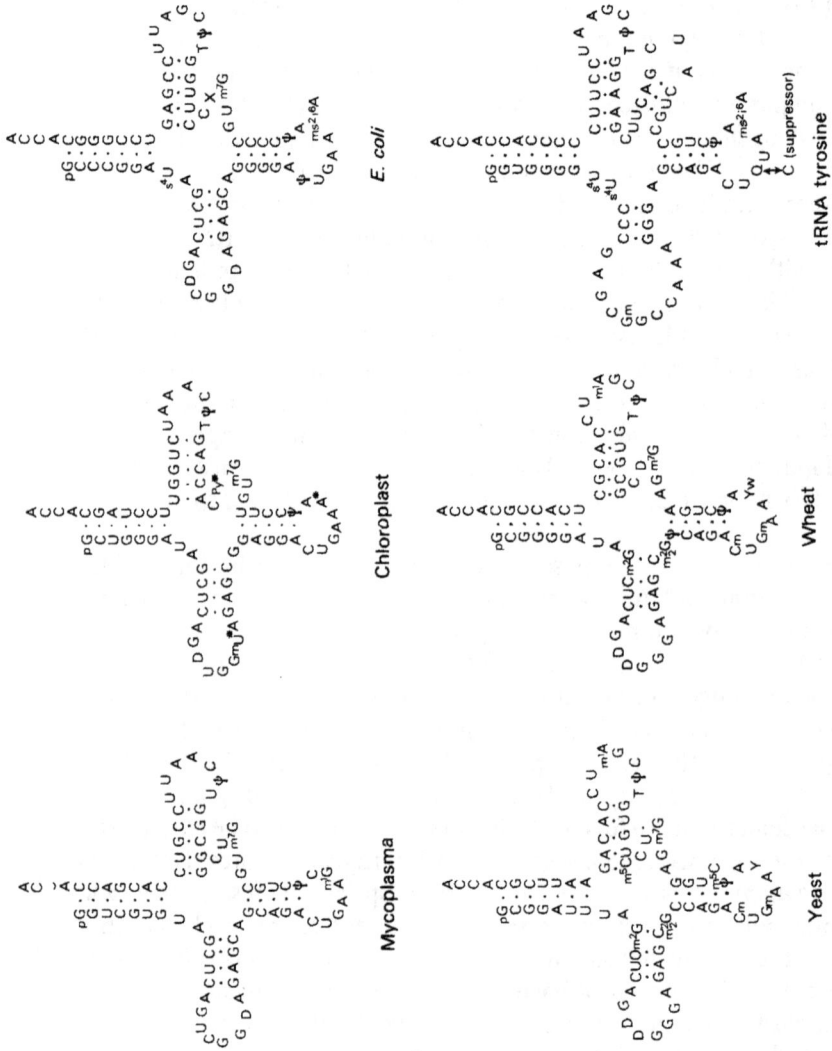

Figure 6   Transfer RNAs. These fold up in a cloverleaf arrangement which includes three, sometimes four, loops. The anticodon is formed by the three middle bases of the lower loop. The sequences of five phenylalanine-specific tRNAs from very different organisms are represented here together with the sequence of a nonsense suppressor tRNA specific for tyrosine (*see* Chapter 9). Some mitochondrial tRNAs have only two loops.

even 90°C, the acidophiles which require an external pH of 2 or less, the halophiles which need the high concentrations of salt that occur in the Dead Sea or the Great Salt Lake. If we take any two of the three kingdoms defined by Woese, we find a number of common features. Thus, most tRNAs of standard bacteria and eukaryotes contain the modified bases thymine and dihydrouracil, and the two are virtually absent in archaebacteria. The membrane lipids of both standard bacteria and eukaryotes are made of straight ester-lipid chains, while those of archaebacteria are branched, ether-lipid chains. Now, archaebacteria and standard bacteria are similar in size, are devoid of nuclei and organelles, and their ribosomal RNA fold according to the same pattern. By still other features (the ribosomal protein sequences, the initiation of protein synthesis, the sensitivity of the translation apparatus to drugs like chloramphenicol and kanamycin), archaebacteria and eukaryotes resemble each other and are distinguished from standard bacteria.

Besides proteins and nucleic acids, other molecules have a message to give on the ways of evolution. Hormones, made of short peptides, show sequence homologies despite wide differences in function. And lipids are the only conserved large molecules in fossils. . . .

4 Evolution in three dimensions

Proteins initially seem to be compact masses in which the amino acids are tightly packed together, leaving no space for water molecules to circulate. Removing the amino-acid side chains reduces the protein to its skeleton, the turns and folds of which do not seem to obey any logic. Gradually, however, we are learning to see proteins in space. A long education in observation was necessary before these disordered bundles became intelligible, appearing as simple combinations of basic structural patterns with a defined repertoire of zig-zags and turns. The first major progress towards deciphering protein structures was the discovery by Pauling and Corey of three conformations, which are still universally recognized as the basic structural arrangements in proteins: the α helix, the parallel β pleated sheet and the antiparallel β pleated sheet (Fig. 7). After this, 20 years were spent understanding how these elementary conformations were combined to form higher order structures.

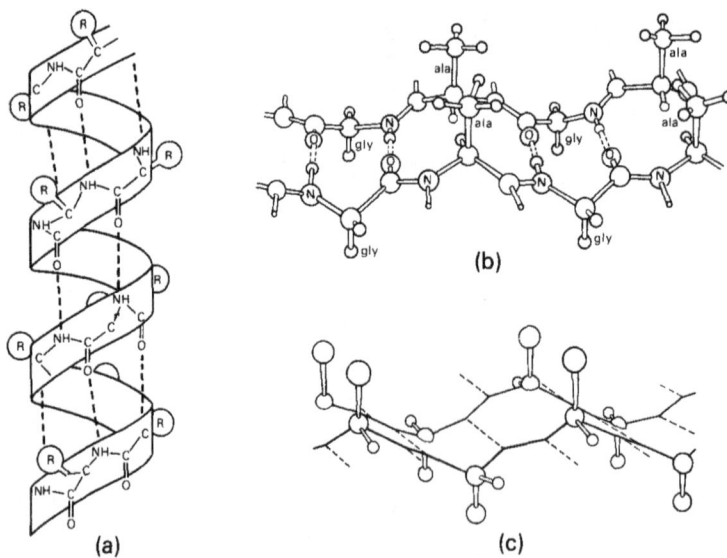

Figure 7 α *and* β *conformations.* (a) The stability of the α helix is ensured partly by interactions (symbolized here by dotted lines) between each amino acid and one situated one turn further on in the helix. (b) and (c) A β pleated sheet.

Amino acid chains show quite a strong tendency – locally at least – to adopt either a helical conformation or a zigzag conformation. Two zigzags side by side form a pleated sheet which is called parallel or antiparallel according to whether the two chains progress in the same direction or not (Fig. 7). Certain amino acids favour helices. Glutamic acid, alanine and leucine are often found in α helices. β pleated sheets are rich in methionine, valine and isoleucine. Other amino acids – notably proline, glycine and tyrosine – do not often occur in α or β ordered structures. They are found at turns. We may thus represent a protein chain as a succession of ordered

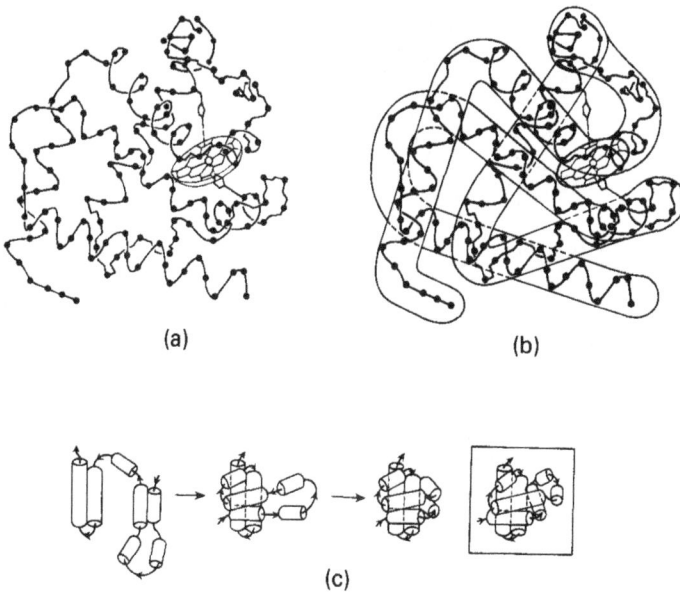

(a)   (b)

(c)

Figure 8   Structure of haemoglobin. By reducing each amino acid to a point as in (a), one can follow the spatial folding of the protein chain. When the amino acids are given their volume, the molecule seems more like a sausage (b). In (c), helical zones of the molecule are represented as cylinders which link non-helical zones. This representation makes it quite easy to understand how the haemoglobin chain attains its definitive folding ((a) and (b) According to Perutz, reference 19; (c) according to Lim and Efimov, reference 176).

regions (α or β) and regions connecting them. Now, two ordered regions which follow each other in a protein sequence are nearly always close neighbours in space which are not linear extensions of each other but lie side by side. The segments joining them thus form loops and several ways of forming the turns are known.

The spatial structures of proteins can be more easily understood if they are represented with α helical regions as cyclinders, β zigzags as flat arrows and the turns as wires (Figs. 8(c) and 9). When we start with an unwound protein chain and let it fold up, it seems that at first certain zones contract to form zigzags or helices. These conformations then regroup to form a more compact mass which we might envisage *a priori* to be one of many possible types. But here again we find higher order conformations appearing. One of the most common arrangements consists of the succession of three ordered regions: β zigzag, α helix and β zigzag, with the two zigzags associating to form a parallel β pleated sheet. The α helix can connect the two zigzags either by passing below the sheet or above it. In 98% of the cases listed, the connection is made

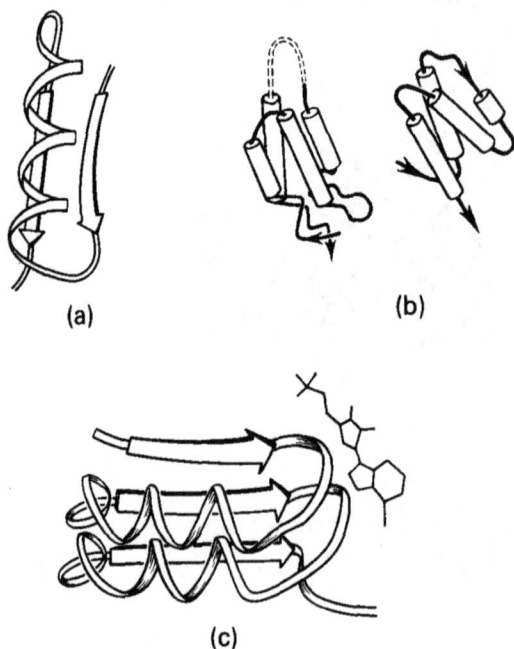

(a)   (b)

(c)

Figure 9   Higher order conformations and domains. Here β zigzags are represented by flat arrows. In the β–α–β structure, the α helical section which links the two β zigzags almost always passes as shown in front and not behind. The structure with four α segments is common to a viral protein (TYMV) and an activating enzyme for tyrosine. The nucleotide fold (c) is found in nearly all proteins which use nucleotide derivatives as auxiliary reactants, including activating enzymes. See references 60, 223, 281.

at one side (Fig. 9a). Another regular feature is the tendency of β pleated sheets to stack and form double layers of β pleated sheets. The ordered regions in haemoglobin are exclusively α helical. We therefore represent them as a necklace of cylinders joined by wires. The number of ways in which cylinders can be grouped together to form a compact globular structure is rather limited. Figure 8 shows how, with this type of consideration, the structure of haemoglobin appears to obey an extremely simple logic.

Levitt and Chothia place globular proteins into four categories: all α, all β,

α/β and $\alpha + \beta$. In the first the only ordered regions are α helices. Proteins in the second category contain predominantly β pleated sheets, which are always present in double layers. In category $\alpha + \beta$, helices and zigzags are both present but in separate parts of the protein; while in class α/β, helices and zigzags alternate.

Armed with these concepts, we can compare the three-dimensional structures of proteins. We can distinguish various super-conformations, including the nucleotide fold present in all proteins which bind nucleotide derivatives (for example, NADH-linked enzymes), and the immunoglobin fold. Astonishing analogies have been found between proteins of seemingly unrelated function and sequence. For example, the terminal lobe of anti-

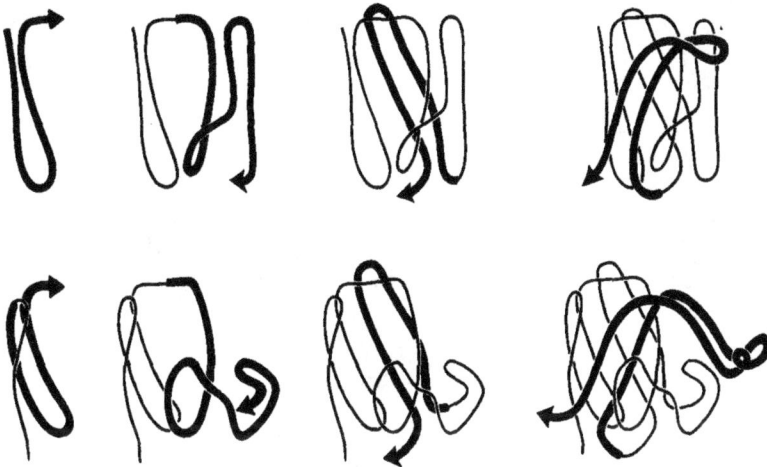

Figure 10   Foldings. Similarity between the spatial folding of an enzyme, superoxide dismutase (down), and an antibody chain (up), according to Richardson (reference 225). This drawing reveals the two structures progressively by successive additions shown as thick lines.

bodies has a structure very similar to that of an enzyme catalysing an oxido-reduction reaction: superoxide dismutase (*see* Fig. 10). In other cases, a striking similarity of spatial structure is corroborated by less-clear similarities of sequence: the best example is that of lysozyme and lactalbumin. Very large proteins can be resolved into a string of globules. The heavy chain of antibodies (Chapter 14) is formed of four globules; the light chain contains two. Four of the six globules have closely similar spatial structures and show similarities at the sequences level (Fig. 11). This suggests, as for haptoglobin $\alpha2$, that the sequences could have elongated after gene doubling and fusion. Similarly, a protease, endothiapepsin, seems to be formed from two globules of closely similar spatial structure (Fig. 11).

The three-dimensional shape of proteins thus seems to be dictated by simple principles of folding which lead to a limited number of combinations.

The possibilities for the formation of the first ordered stretches are more limited than one might have thought. A tendency to form ordered zones is shown in polymers (synthesized in the laboratory) of alternating D- and L-amino acids, as shown by the elegant analyses perfomed by Spach's group. The amino acid L-benzylglutamic acid polymerizes to form an α helical polypeptide. The polypeptide formed from the D- isomer of this amino acid also has a helical structure which is the mirror image of the first. A copolymer of strictly alternating D and L forms of the amino acid might be expected to form frightful zigzags, with the chain constantly changing direction. In fact nothing of the sort happens; these alternating polymers can form several

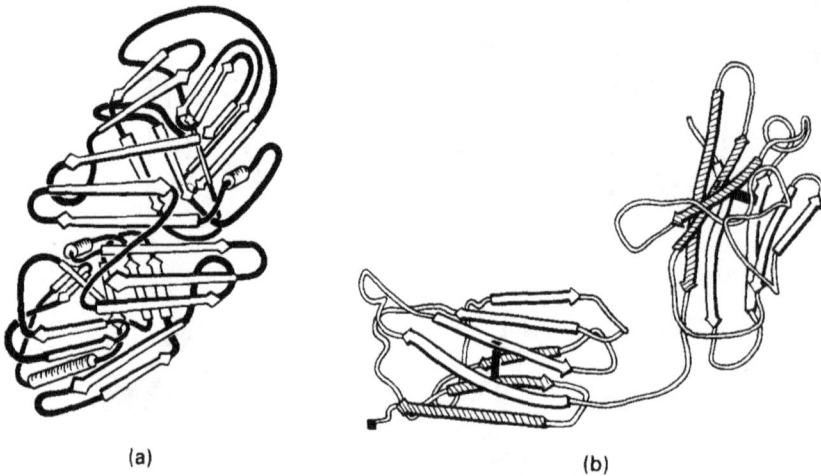

(a)   (b)

Figure 11   Globules. Large proteins can contain several globules. Sometimes two globules in the same protein show striking similarities in their structures in space. Antibodies are formed from two chains which form six globules in all, four of which have very similar structures (b). Note the arrangement of β regions in 'barrels'. A degradative enzyme, shown in (a), shows two similar globules. See references 235, 251, 182.

regular helical structures. One of the conformations adopted is the α helix which remains α helical for several residues, is then interrupted, and starts off again in the opposition direction for several turns. A still more surprising structure is the π DL double helix. The alternating D–L copolymers form a double helix in which the two chains are equivalent, and which can extend indefinitely.

Once a polypeptide folding which allows a given catalytic activity to be performed has been obtained during evolution, it seems that improvements in its specificity are acquired not by radical remodelling of the active site but by fine modifications. In particular, adding an appendage to the protein can cause certain zones of the active site to be compressed or relaxed. Proteins often contain more than one polypeptide chain and the attachment of a regulatory substrate to one of the chains governs the catalytic activity of another. When a protein is made of several structural domains, these can usually fold independently (but there are exceptions in small proteins). Separate domains in eukaryotic proteins often, but not always, correspond to separate exons in their structural genes, and Gilbert proposed that the exon–intron system could form the basis of a combinatorial kit, allowing many different proteins to be built out of the same parts. However, when a large protein is composed of two domains, this can be usually traced back to an internal duplication event. According to the general view outlined here (the question will be taken up again in Chapter 12), the formation of protein structures which might be reasonably interesting for living processes appears relatively easy, since structures can be combined in only a limited number of ways. Having produced a first prototype, evolution can later remodel it to improve its catalytic activity and refine its regulatory potential.

But one can support an opposing view, namely that most of the effort in perfecting sequences has gone into stabilizing the three-dimensional structure. It is highly plausible that not one but several spatial arrangements correspond to a given sequence and that evolution occurs to reduce the range of possibilities. In this case, one strategy would be to produce the stablest possible standard pieces (α helices, β pleated sheets), then optimize the packing between these pieces, through tight hydrophobic contacts (this will restrict the possible angles between two α-helices in contact to values around 20, -60 and $-80°$, and also impose constraints on the packing of β sheets). The same solutions could very well have been found independently on several occasions for several different families of proteins. Alternatively, having realized that (at least for a primitive protein) the three-dimensional structure is not necessarily unique, it is possible to imagine courses in evolution in which a superfluous protein can, after mutation, switch to a new spatial arrangement. This proposal was made for ferredoxin and more and more attention may be paid to this type of possibility.

5 Can sequences be compared?

We have been able to exploit the enormous amount of work on determining protein structure in two ways. First, we compared sequences to deduce their family relationships and ancestry in a neo-Darwinian perspective: genes mutate, duplicate and mutate again. Sequences, initially present in low number, diversified through the ages and thus gave rise to the present immense variety of sequences. This is divergent evolution. Later, we compared the three-dimensional structures of various proteins and worked out some laws: the universality of certain structural patterns, restrictions on ways of combining these in higher order arrangements and the preferential use of certain amino acids in each conformation. Looked at from this point of view, protein structures are perfected little by little by improving the stability of the basic conformations and searching for the most favourable combinations of these conformations. Starting with a great diversity of sequences, evolution leads to a limited number of standard structures. This is convergent evolution. If we detect homologies between distantly related sequences, it will be difficult to decide between the divergent interpretation (both sequences descending from a common ancestor) and the convergent one. Let us put these studies on a more concrete basis and compare fragments of sequences, for example:

```
. . . . . . B │L A│K E│M I R│E . . . . .
. . . . . . D │L A│T H│M I R│G . . . . .
```

Out of ten pairs of amino acids there are five which correspond. Is this important? If, for simplicity, all amino acids occur with equal probability, the chance of having an A below an A or an M below an M is one twentieth for each amino acid examined. Knowing this elementary probability (1/20) we can calculate the probability of finding five coincidences in two segments nine residues long, taken at random as above. If there are 'structural constraints' in the sequences which have been compared, the calculated probability becomes meaningless. Suppose that the fragments of sequence which have been compared form, in their respective proteins, a structural pattern which requires strict alternation between hydrophobic and hydrophilic amino acids. Suppose that here we designate the hydrophilic ones by the letters A to J and the hydrophobic ones by the letters K to T. Under these conditions, as soon as we have placed the hydrophilic D below hydrophilic B, the fragments are placed in register so that we will always find a hydrophilic amino acid below a hydrophilic one and a hydrophobic one below another. Then the probability of finding M below M is no longer 1/20 but 1/10.

Of course, if we know the structural constraints we can take them into account and correct the probability estimates. But, unfortunately, certain invisible constraints are quite often capable of being reflected in subtle and unforeseen ways in sequence homologies. Internal repeats, inversions, local constraints on composition, palindromes and every deviation from the random model can artificially increase the coincidences. Conversely, the rules governing the presence of amino acids in each type of structure lose their validity if they have been established from proteins with common ancestors. The presence of a particular amino acid at a particular position, which might suggest a structural requirement, could be merely a vestige of ancient but incompletely dissolved ties of kinship. Proteins which appear totally unrelated can conceal subtle 'constraints of phylogeny', if, for example, they come from ancestral proteins whose genes were complementary or overlapped with one base shift.

Let us forget these criticisms and see how we can deduce the evolutionary relationships between haemoglobins from comparisons of their sequences. The two trees in Fig. 5 were constructed using two different approaches. We notice in the first (produce by Dayhoff and Eck) that the haemoglobin β chains have evolved much faster than the α chains, while the second (Goodman, Moore and Matsuda) gives the opposite result. First, let us define the 'distance' between two sequences.

Having superimposed two sequences (as in Table 2, p. 18) we can count the number of positions where the upper amino acid is not the same as the one below. This number provides an index of the degree of differences between the two sequences and can thus measure the 'distance' between them. Working along these lines, we assume that all amino acid substitutions are equivalent. In other words we count one point when a valine is replaced by an isoleucine, which resembles it closely, or by a histidine which does not resemble it at all. To overcome this drawback methods of quantifying differences more finely have been sought. Most often these are guided by the genetic code. One point is counted for amino acid substitutions which can occur by changing one base in a codon, two points when at least two bases must be changed, and three when all three bases have to be changed. This can be refined by taking into account the incidence of each type of mutation ($G \rightarrow A, G \rightarrow T$ etc.) in DNA. But we can also use a functional weighting: if an amino acid is replaced by another of similar properties we consider that the sequences have diverged very little. If, on the contrary, it is replaced by a very different amino acid, the substitution will carry more weight in the measure of distance. To evaluate the weight of each type of substitution, we can use empirical indices of similarity between amino acids or, more simply, use the table of all amino acid substitutions observed in protein families. We can assume that if a substitution is common, it is easy to perform and thus represents a very small evolutionary step. We must therefore give it a small weighting. All this is perfectly logical and obvious. However, in my opinion, the reverse choices are equally well justified. The aim of these measures of distance is really to allow evaluation of the time elapsed since the proteins diverged from a common ancestor. A radical change of amino acid, apt to greatly modify the properties of the protein, will cause rapid selection or counter-selection. A very conservative change, which would procure minimal selective advantage, would take a long time to become fixed in the species. Thus there might be an inverse relationship between elementary measures of morphological distance and time intervals. Perhaps in the distant future the constructors of phylogenetic trees will take note of this remark. The present climate does not favour critical reflection.

The distance between two sequences is the sum of elementary distances evaluated for each amino acid substitution, possibly modified by a correction coefficient for large distances. If two sequences 100 residues long differ at eighty positions, about four-fifths of the mutations occurring at random along the chains will leave unchanged the twenty positions where the amino acids correspond and will not increase the distance between the sequences. Thus going from eighty to eighty-one differences between the sequences represents on average not one but four or five mutagenic events. The correction factor which is used is linked to an underlying model of evolution. Usually, evolution of sequences by random mutations is simulated in a computer, with the assumption that mutations accumulate at the same rate in lines which have diverged. This gives the relationship between numbers of mutagenic events and numbers of substitutions.

We thus have a set of distances between proteins in the same family. This allows us to represent them like towns on a road map in which each road has the distance between two towns marked on it. Instead of triangulating in this way, our real objective is to place the proteins at the extremities of a phylogenetic tree. Distances between present-day sequences are only used to reveal kinship and to evaluate the time elapsed along each branch of the tree. In other words to use the direct speech of academics, we want to put synchronous relations into a diachronous context. There are several methods for transforming the map of distances into a phylogenetic tree (*see* Fig. 12). Once the tree is constructed, we can read on it the evolutionary history of the protein. This tree can in its turn be considered as a road map to which have been added some towns (the ancestral proteins at the branch points of the tree) and roads, while others have been removed. A new town will be, for example, 18, 20 and 15 distance units from three old towns. However, there is nothing to guarantee that we can find a protein sequence which is simultaneously at these three distances from three initial sequences. This difficulty, and others, were underlined by Vogel and by Beyer and collaborators.

Another criticism of this approach is that it reduces all the information about sequences to numbers – the distances – which is an impoverishment. We could get by without knowing the sequences at all, since differences between proteins of unknown structure can be estimated immunologically (this is explained in Chapter 15). Most constructors of trees prefer, like Fitch, to analyse ancestral relationships using sequences, position by position. These reconstructions are generally guided by the Principal of Parsimony, according to which evolution always takes the shortest route from one sequence to another. So we look for ancestral sequences which could have given rise to present-day sequences with a minimum of mutations. This method, when applied to haemoglobins, gives the tree shown in Figure 5(b), which we shall now examine.

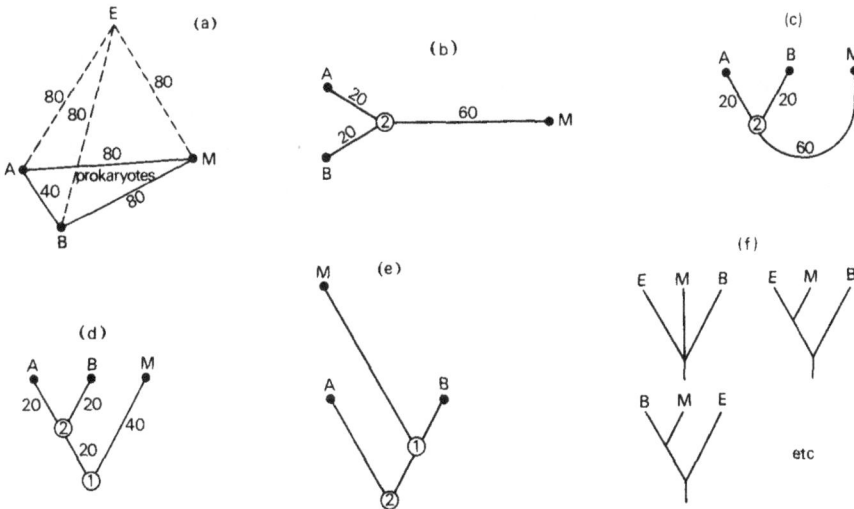

Figure 12 Constructing trees. Here it is assumed that the 'distances' have been measured between sequences from an ordinary bacterium (B), a eukaryotic organism (E), a methanogenic bacterium (M) and a blue-green alga (A) and that the relations shown in (a) have been found. Construction of a family tree (d) which results from analysis of the results compiled in (a) passes through steps (b) and (c). Note the crucial transformation of (c) into (d): the branch-point N° 1 has been introduced half-way between M on the one hand and A and B on the other. By placing it arbitrarily in this position, we imply that the sequences have evolved at the same speeds along the two branches which start at 1. By rejecting this hypothesis, the results can be represented, taking into account possible fluctuations, by a tree like (e). Similarly, E, M and B can be grouped in various ways (f).

The tree gives an account of the evolution of all the globins from an ancestral globin A and intermediates B and C. The common ancestor of all α and β haemoglobins is C; B is the ancestor of C and the myoglobins, and lamprey globin and B are derived from A. The distance between A and present-day chains are quite variable. We find:

A–lamprey globin   : 144 units (mean of two sequences)
A–myoglobins   : 240 units (mean of nine sequences)
A–β haemoglobins : 268 units (mean of twenty-one sequences)
A–α haemoglobins : 287 units (mean of eighteen sequences)

The rate of evolution is not at all the same in the various branches of the tree. The explanatory scenario of Goodman, Moore and Matsuda is as follows: at some point in time gene A doubles. One copy of the gene continues to perform its usual role, for which the globin sequence is already well adapted so that it has no need to vary. The second copy (B) specializes in a completely new function. At first it evolves a great deal to adjust to its new role; once gene B has adapted to the role of myoglobin, its evolution slows down. But now B doubles in its turn. One copy continues to control myoglobin production and evolves at a slow pace while the copy which leads to the haemoglobins mutates at full speed until it stabilizes. We should note that differences in rates of evolution are even more important than would appear from the comparison of sequences given above. All variation in speeds of evolution is concentrated in the time intervals when a protein is in the process of changing function. At the end of such intervals, all speeds return to normal. The fact that normal rates of evolution are roughly equal was foreseeable. If we count the differences between the globins of man and horse we find:

human myoglobin–horse myoglobin : 19 differences
human β chain–horse β chain   : 25 differences
human α chain–horse α chain   : 19 differences

These crude data clearly suggest that from the moment when man and the horse diverged, myoglobin and the α and β haemoglobins evolved at roughly the same speed. The scenario imagined by Goodman, Moore and Matsuda, very reasonable and coherent though it is, collapses when it is examined in the light of the following idea. Phylogenetic tress constructed according to the parsimony principle produce branches which are the longer the more ramified they are. There is a quite strong correlation between the distance given above and the number of sequences from which the averages were taken. In the tree produced by Eck and Dayhoff, β haemoglobin clearly evolves faster than α haemoglobin – but in this tree there are twice as many β-type sequences as α-type sequences! The phenomenon of elongation of very ramified branches is an easily understood artefact of the techniques of constructing trees. If we want just to connect two points in empty surroundings, the most parsimonious path is the straight line. Where there is a good deal of information, the path has to be lengthened so as to branch off near neighbouring points.

Where there is a good deal of information, the path has to be lengthened so as to include every point. If more-balanced numbers of sequences were used, there should be more-equal rates of evolution. This would not prove that the equality revealed in this way was real. In fact, if there were truly unequal evolution along two great branches of a tree, present techniques for reconstructing phylogenies would tend to obscure rather than magnify the inequalities.

If we constructed a phylogenetic tree from twenty known sequences, and we then add two newly elucidated sequences, will the tree be profoundly modified? If the answer is yes, the construction technique produces 'unstable' trees which can be called into question. If the trees are always stable, this means that the technique does not make the best use of the information. An error will never be corrected. Preference for one technique or another is finally a matter of taste. This area of molecular evolution is steeped in empiricism.

6 Replication and genetic tinkering

By replicating, DNA remains indefinitely like itself. Its invariability is both real and deceptive. When a pure strain of bacteria is cultivated in the laboratory, where it is well isolated, its DNA is reproduced without great changes over many generations. Of the five million base pairs in a bacterial DNA, less than one is changed on average per cell division. DNA is constantly under attack: by ultraviolet rays which induce cross-links between neighbouring thymine residues or by the action of various cellular molecules which can cause breaks in the sequence. Cytosine is slowly and spontaneously transformed into uracil. Enzymes correct the malformations and restore the DNA. The pieces of broken DNA are not always rejoined in the correct order and sometimes a fragment of foreign DNA (from a virus, for example) is integrated into the repaired DNA. Thus enzymes responsible for maintaining DNA molecules sometimes contribute to chromosome remodelling.

Let us describe replication at the most elementary level, in a test-tube mixture of template, the four replication precursors (dATP, dGTP, dCTP, dTTP) and DNA polymerase. A nucleic-acid chain is a chemical object which moves and deforms. The double helix is breathing: chains dissociate along part of their length and reassociate. The bases in a chain can move apart, turn around and present new possibilities for pairing, like the letter M which becomes W when it is inverted. Atoms can migrate within a base so that its chemical structure oscillates between two forms called tautomers. Adenine spends a fraction of its time (about one hundred thousandth) in a state resembling guanine. The text of an isolated DNA chain fluctuates and changes its meaning. If we took a snapshot of the molecule one base in 100 000, perhaps more, would appear different from its usual shape. Replication of DNA by enzymes in the cell is even more faithful than the best photographs, since the error rate is about one in 100 million.

A DNA polymerase copies the DNA at full speed, making mistakes, and other enzymes correct them later. Sometimes the polymerization and repair activities are performed by the same enzyme. These DNA polymerases can elongate a chain by adding a base or shorten it by excising the last base incorporated. The two activities compete. When the terminal base is complementary to the one facing it, the excision is not very effective, so there is a strong chance that the enzyme will incorporate a further base and advance one notch. When the terminal base is 'incorrect', the excision is efficient and generally occurs before the enzyme has had time to incorporate the next base. The enzyme then moves back one step and starts again. But sometimes the next base is incorporated before the terminal base can be excised and the enzyme advances one step, letting the error remain. If the mistake is not repaired subsequently it remains, becoming a mutation. In bacteria, as in phages, there are fast-mutating strains, which mutate much more often than the common strains, and slow-mutating strains, in which mutations are rare. In most of the cases studied, the rate of mutations is linked to the level of precision of DNA polymerase function which is itself linked to the relative effectiveness of incorporation and excision. A fast-mutating polymerase excises badly; it continues and does not give itself enough time to correct errors. The slow-mutating one, on the other hand, effectively excises correct bases as well as incorrect ones. It often moves backwards, on average going two steps backwards for three forwards. This means that it consumes, on average, three nucleotides to incorporate one. Precision is expensive energetically.

After synthesis of a DNA chain, methyl groups are added to it at specific positions. A brand new chain is thus bare of methyl groups. A replication error will show up as a mismatch, say between A in the old strand and a G in the new one. When the repair system detects the G.A mismatch, it turns it back to A.T, not to G.C, knowing by the criterion of methylation that the A-carrying chain is the old one, as shown beautifully by Meselson *et al*. In one case at least, it is known that the methylation pattern is built up by a process similar to replication. When the methylase encounters a (Methyl-C)–G sequence, it adds a methyl group on the C of the C–G sequence facing it on the opposite strand. Defects in this process are common, and give rise to 'paramutations'.

For a long time mutations were mysterious. What was the origin of these abrupt changes? It used to be thought that, in addition to the normal replication processes, there were special mechanisms in the cell for producing mutations. Without variation there would be no evolution. So the development in the history of life of mechanisms for giving rise to mutations was believed to have been one of evolution's most brilliant inventions. This is an enormous misapprehension. On the contrary, evolution's great achievement was the acquisition of accuracy in replication and genetic translation. Increases in complexity are conditioned by improvements in precision. A child's scooter which has oval wheels, a bendy frame and loose nuts will run, but for a supersonic plane to function its essential parts have to be machine-finished and assembled with the greatest care.

Let us now describe replication in a cell, the bacterium *E. coli*. The DNA is circular, without free ends, while maintaining its double helical structure. The two chains thus form two circles twisted across each other so that it is impossible to seperate them unless at least one of them is cut. To replicate bacterial DNA, enzymes cut one of the circles, allow it to unwind and close it again. Strands complementary to the two circular chains are made as the coiling of one circle around the other is relaxed. Finally if all goes well, two separate double-circular chains are obtained. Sometimes, however, the DNA molecules fail to separate and stay together, like the links in a chain. This replication accident is often observed with the circular DNAs of mitochondria and phages. Knots are sometimes observed in phage DNAs. The number of turns which one circle makes around its partner is not left to chance. There are enzymes which cut the circles and rejoin them while controlling the state of twisting of the DNA. To a large extent, cellular regulation depends on interactions between regulatory proteins and specific sites on DNA. An over- or under-coiled DNA molecule would impede the regulatory devices. Circular DNA is stable, in that one region cannot be unwound without causing increased twisting on either side of it. Other advantages of circularity have been suggested: no free ends, hence less vulnerability to attack and greater flexibility in regrouping genes. A form intermediate between the 'open' double chain and circular DNA is found in certain phages.

DNA chains are oriented. The chemical bridge between two bases is not symmetrical but links the 3' position of one nucleotide to the 5' position of the following one. This succession of 3'–5' bonds defines the direction of reading along the strand. During replication, nucleic acids always elongate in this direction. Since both strands of the double helix are copied at the same time and they are associated head to tail, the polymerases which copy them move in opposite direction thus:

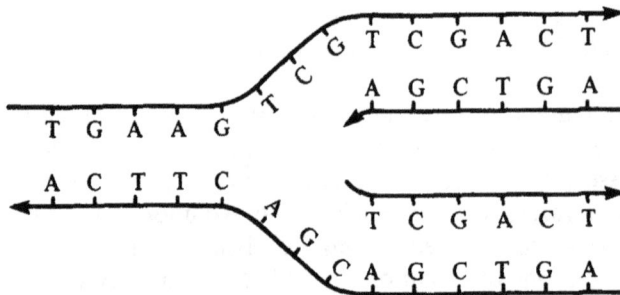

Overall, however, replication progresses in a well-defined direction. In the gene sequence P,Q,R,S,T,U,V etc., Q is replicated before S and S before U. But during replication of zone Q, one of the strands of the new double helix is synthesized in the direction P → R and the other in the direction R → P. The DNA is replicated in small portions which succeed one another along the chromosome and which have to be connected afterwards. The spurt of replication is complicated. DNA polymerases add nucleotides onto pre-existing chains adequately but cannot easily perform the first step when there is no primer. In *E. coli* it seems that DNA replication starts with the synthesis of a short RNA chain of about 100 nucleotides. Next, the DNA polymerase joins on a thousand or so deoxynucleotides. The RNA primer is then degraded. A DNA chain from a bit further along, bearing a length of RNA, is extended to rejoin the first, whose primer has disappeared. The two DNA segments are then joined by a DNA ligase to make one chain. The same cycle of events starts again further along. Could the inclusion of RNA in DNA replication be a vestige of a time when the genetic material might have been an RNA double helix?

All sexually reproducing organisms possess chromosomes which occur in pairs. The individual obtains one complete set of chromosomes from the mother and one from the father. Each chromosome can undergo intimate exchanges of hereditary material with its partner by a procedure which is not yet fully understood. This is genetic recombination, which ensures permanent intermixing of genes, remodels DNA and induces repetition of genes. During recombination the parental chromosomes pair up along their entire length with homologous genes facing one another:

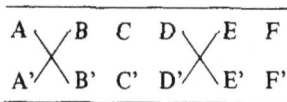

$$
\begin{array}{cccccc}
A & B & C & D & E & F \\
A' & B' & C' & D' & E' & F'
\end{array}
$$

An event called crossing over can occur at any point of contact between the two chromosomes, symbolized here by an X. After the two crossings over shown above, the genes are redistributed and we end up with the reorganized chromosomes A B' C' D' E F and A' B C D E' F'. More precisely, the exchanges consist of transfer of DNA between double helices; thus

```
A A T T G C G C T A          A A T T G C G C T A
T T A A C G C G A T          T T A A A G C G A T
        ×            becomes
A A T T T C G C T A          A A T T T C G C T A
T T A A A G C G A T          T T A A C G C G A T
```

The pairs G.A and T.C are not complementary. Thus we pass through a stage in which each of the two new double helices contains an imperfection. If these imperfections are not repaired, the pairs G.A and T.C will give rise in the next replication cycle to the pairs G.C, T.A, T.A and G.C so that the two *original sequences are re-formed.* But sometimes an imperfection is detected and the pair G.A is repaired to G.C or T.A. Which is the correct choice? There is a case in which the answer seems obvious. If there is a deletion in one of the DNA molecules (a piece of sequence is missing) it might be decided that the incorrect molecule is the shorter of the two. Bernstein suggests that the primary objective of recombination might be to detect deletions and fill in the gaps.

Recombination makes use of enzymes which can cut DNA and rejoin the pieces. Sometimes a length of chromosome is joined back to front. The chromosome A*BC*DEF becomes A*CB*DEF. If the messenger RNAs of genes A and B were initially transcribed from the same DNA chain, then after inversion they will be transcribed from opposite chains:

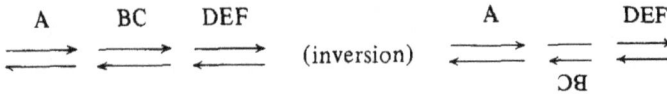

Here I have represented the two DNA chains before and after inversion of the central segment. Two different genes, read in opposite directions, can occupy a common section of DNA thus:

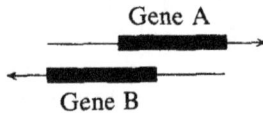

This example is taken from phage λ. Other Genes are known which overlap on the same chain of phage DNA. The base sequence ACGCTGGACTTTGTGG is read for one gene as the codon series ACG, CTG, GAC, TTT, GTG, etc., and for the other gene as CGC, TGG, ACT, TTG, TGG, etc. Overlapping genes are believed to be confined to phages and perhaps other viruses. The reverse situation applies in higher organisms; the codon series of a gene on the DNA can be interrupted by non-codon sequences which are removed after transcription. Genes may thus be split into two, three or more sections separated by tens or perhaps hundreds of nucleotides.

When one of a pair of complementary chromosomes contains an inversion with respect to the other, pairing still takes place but it necessitates a special configuration which entails risks: sometimes during cutting and rejoining, one of the chromosomes recovers two copies of the inverted segment while the other ends up with none. This effect can be amplified, two copies becoming four, etc. Chromosome inversions are a source of genetic instability. Other chromosomal accidents, which I shall not describe, can also lead to the duplication of certain regions. When the cellular reproduction cycle is in motion, chromosomes can sometimes be duplicated without the cell dividing so that it then possesses a double set of chromosomes. After several adjustments, a stable line is sometimes established (in plants) giving a new species having twice as many chromosomes as its ancestor. Two cells from different species may be combined to form a hybrid cell with a set of chromosomes from each of the parent species. Fertile hybrids are known in plants; and some viruses, bacteriophages, can remove a piece of chromosome from an organism then integrate it into a chromosome of another organism. This achieves genetic mixing between species which would otherwise be incapable of crossing. What is the role in the evolution of natural populations of such gene exchanges through viral intermediaries? We are hardly in a position to evaluate it. But the importance which is attached to this possibility is growing, doubtless because scientists have known for only a short time how to perform in the laboratory the sort of genetic tinkering which occurs in natural populations. But the term tinkering is inappropriate, because nature tinkers without an established plan, whereas the laboratory worker knows in advance which gene he wants to transpose. When it is practised by the sages of the Pasteur Institute, genetic tinkering becomes genetic craft, future achievements in which will change the course of human history. Thus famine will be abolished when, having transferred ruminant genes to himself, man will be able to feed himself directly by grazing in the meadows. Chimpanzees made intelligent and docile will provide cheap and non-unionized manual labour. We should also remember the German project, reported by de Pawlowski, to raise beer-producing cows. The Americans, always ahead, propose marketing the sense-of-humour gene thus relieving the thirty-five million U.S. citizens who, according to the latest biometric surveys, appear to be deprived of it.

We can make sequences evolve in the test tube. Spiegelman replicates a viral RNA with the RNA polymerase from the same virus (Qβ). He removes a small fraction of the replication product and uses it as the starting point for a second series of *in vitro* replications. This cycle of operations is repeated 1600 times. If variants of the initial RNA are formed in the course of a cycle which are replicated more easily than the other RNA molecules, their proportion must increase with each cycle. The end result is that the initial viral RNA is supplanted by the more rapidly replicating molecules. The victor in these experiments in artificial molecular selection or 'extra-cellular Darwinism' is a rather puny specimen: a tiny piece of RNA, ninety nucleotides long. We might have expected this. The RNA polymerase takes less time to reproduce this mini-monster than the initial viral RNA, which is fifty times longer. This type of experiment, which appeared very promising at the time, has led to just one result of evolutionary interest. At high concentration, as found by Sumper, the enzyme is capable of synthesizing, in the absence of any contaminating template, a well-defined piece of nucleic acid with a sequence closely related to those of some previously known mini-monsters. It seems that if several nucleotides, each carried by a molecule of RNA polymerase, are brought into close contact, they form an acceptable short template for another molecule of RNA polymerase. Once short oligomers are present, the synthesis of longer ones is made easier.

DNA sequences often form palindromes, the name given by biologists to sequences which are almost perfectly complementary to themselves. GCAG-CATATGCTGC is one. To see why let us write the complementary sequence underneath:

```
1——G C A G C A T A T G C T G C——→
 ←——C G T C G T A T A C G A C G——2
```

Sequences 1 and 2 are identical. In isolation, they can fold up on themselves in a hairpin structure:

```
———— G  C  A  G  C  A  T
                       ⌉
———— C  G  T  C  G  T  A ⌋
```

Palindromes may occupy strategic positions in DNA. Most enzymes which cut DNA at specific sites do so preferentially in palindromic sequences. Palindromes have been invoked to explain the choice of sites for genetic

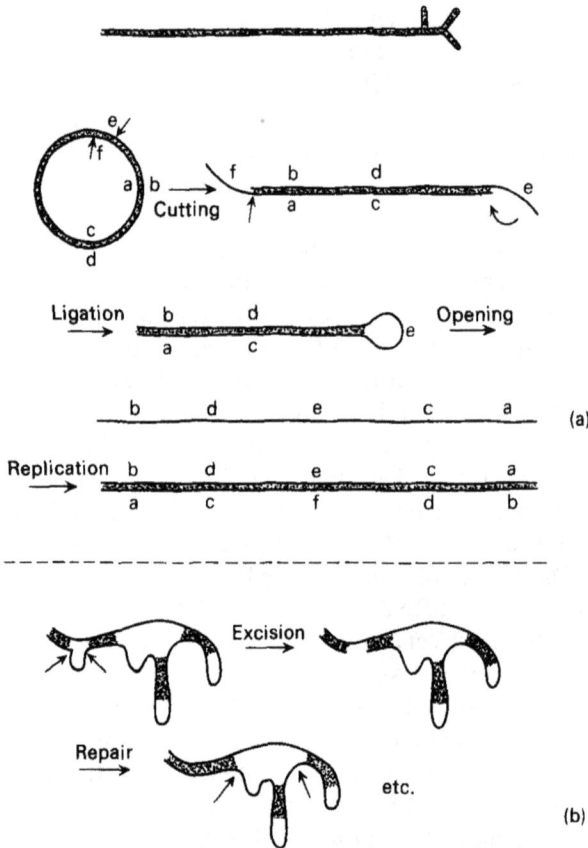

Figure 13   Palindromes and re-sorting (or tinkering). A DNA chain bearing ribosomal RNA genes is capable of pairing with itself to form a very long hairpin with irregular ends. In classical genetics, no mechanism is known which gives rise to palindromic molecules. But, by combining known enzymatic activities, we can explain the formation of palindromic sequences. For example, we start in (a) with a circular double chain of DNA. Enzymes cause breaks and the result is a double chain with single-stranded ends. Then a ligation enzyme can link two ends and transform the previous structure into a hairpin. Finally, replication of this molecule results in the formation of a double chain of DNA twice as long as the initial chain.

recombination, the initiation of transcription and for the attachment of regulatory proteins. Their role is not completely clear. DNA in higher organisms could contain about 5–10 per cent palindromic sequences. What is the genetic mechanisms which gives rise to palindromes? We do not know. Palindromes have still never been produced in the laboratory, although it has been possible to cause duplications or inversions of genes. In principle, the problem is not insoluble, since it should be possible to start with any DNA sequence and, by carefully combining the activities of the range of enzymes which act on DNA, remodel the sequence and end up with a palindrome (Fig. 13).

7 Populations

The idea of evolution by natural selection has arisen several times in the history of human thought. First mentioned by Lucretius, it was taken up again during the last century by Chambers and Matthews, then officially developed by Wallace and Darwin. It is Darwin's name above all which we associate with it. The skill with which he created his public image and attached his name alone to the concept of natural selection is exemplary and inspires many scientists today. The theory put forward in the *Origin of Species* still holds good. Its major difficulties were resolved by combining it with concepts of reproduction taken from genetics. Not only was the fusion of two seemingly contradictory doctrines, Darwinism and Mendelism, possible but it also allowed evolutionists, who had become population geneticists, to found their ideas on the experimental study of populations and to give evolution a decidedly quantitative twist. After the four great founders of quantitative evolution – Haldane, Fisher, Wright and Chetverikov – an immense body of knowledge accumulated whose leading ideas will be outlined here. Evolution-ism is a science. It would be dishonest to reduce its achievements to a clash of key-words: selectionist/neutralist, deterministic/stochastic, chance/necessity.

The simplest situation is that of bacteria grown in a chemostat, a device invented by Monod, Novick and Szilad to follow the evolution of bacterial populations under perfectly controlled conditions. The supply of nutrient at the inlet is regulated and portions of culture medium can be removed so as to leave constant either the total number of bacteria in the chemostat (which is what we will assume below) or the volume of liquid in which they are growing. Bacteria multiply by simple cell division, one mother cell becoming a pair of daughter cells. Suppose we start with 5000 type A bacteria which double every hour and 5000 type B bacteria which double once every 2 hours. After 2 hours, if no liquid has been removed from the chemostat, we will find 10 000 type B bacteria and 20 000 type A. But since the apparatus is set to keep the number of bacteria constant, the proportions 1:2 will be main-tained, giving 3300 bacteria B and 6700 bacteria A, approximately.

The culture eventually becomes enriched in A. Soon B disappears, A having in some way eliminated it completely. The type A has now established itself in the population. The bacteria selected for are those which reproduce faster. There is no struggle between A and B. Instead, the removal of liquid results in a blind selection which affects both bacteria B and A equally. There are few examples in nature of blind selection except great calamities and catastrophes which affect all flora and fauna indiscriminately and are sent to earth to hasten the course of evolution: droughts, famines, floods, earthquakes and volcanic eruptions.

If the population is limited to ten bacteria and the proportions $1:2$ are again reached after 2 hours, there will be 3.33 bacteria B and 6.67 bacteria A, which is nonsense. In reality there will be 3 B and 7 A or 4 B and 6 A. There is necessarily some departure from the exact proportions $1:2$ predicted by the determinist model, and this is not simply because bacteria exist in whole numbers. There are always fluctuations in real processes. When we take a little culture medium from the chemostat, we remove bacteria A and B in proportions which equal, on average only, those in the medium and can fluctuate from one part to another from the mean. When we take this fluctuation into account (that is, if we apply a stochastic treatment), we realize that the outcome of the experiment is no longer certain. The fastest bacteria usually win but the slowest sometimes establish themselves. In small populations elimination of adversaries is rapid but the victor is not always the better.

According to the determinist approach, if all the competitors have the same importance, they should be maintained in constant proportions. The stochastic treatment is more realistic and predicts that one of them will nevertheless end up by imposing itself. For example, in a sexually reproducing population of N individuals an allele appears which at first is present in only one copy and is neither advantageous nor disadvantageous. This allele will have one chance in 2N of eventually being fixed and in this case the complete elimination of its competitors would take, on average, 4N generations. The chance fixation of a character which is not necessarily the most advantageous one is called genetic drift. Studied by Wright as early as the beginnings of population genetics, it is at the centre of present debate – notably since Kimura, who has developed Wright's calculations further, maintained that evolution occurs essentially by drift and fixation of 'neutral' characters which are neither advantageous nor disadvantageous. The neutralists have replaced the true proposition that 'if a newly appeared character is neutral it has some chance of being fixed' with the dubious proposition that 'most characters fixed during evolution were neutral'.

Let us return to the two types of bacteria, A and B, growing in the chemostat. This time A and B grow at the same rate but antagonize each other. Bacteria A excretes a product pA which is poisonous for bacteria B and bacteria B produces an equally efficient poison pB against A. What happens? While bacteria A and B are present in equal numbers the effects of the poisons pA and pB are balanced; A and B are equally affected. But as soon as an important fluctuation occurs, for example a net excess of A, the effects are amplified. The poison pA, produced in excess and acting on a diminished population of B, has more effect while poison pB, produced in smaller amounts and acting on a larger population of A, loses its effectiveness. The imbalance is amplified and B is rapidly eliminated. This biology-fiction scenario has been reinvented on several occasions to explain the origin of asymmetry in the chemistry of living things. Imagine the earth in the past populated with two sorts of cells, the 'left' geobionts with the same molecular asymmetry as today, and the 'right' geobionts with the reverse asymmetry. Both right- and left-life forms develop and evolve in an absolutely equivalent manner for 100 000 centuries. But now each of the geobiont families simultaneously invents a poison capable of eliminating the other. From then on the fate of half the living creatures on the planet is at the mercy of the first major fluctuation. This gives advantage to the geobionts of the left, the effect is amplified, and the right is eliminated completely. Or we can transpose the argument to the molecular level. In the prebiotic soup two families of molecules of opposite asymmetry are formed spontaneously in equal quantities but each inhibits the synthesis of the other. So here we have the reason for the occurrence of L-amino acids in proteins. The beauty of this argument fades when one realizes that the assumptions of the second part contradict the first. If fluctuations are to be taken into account, as they are at the end, then they must also intervene at the beginning. In this case there is not the slightest reason to think that two life forms, one 'left' and the other 'right', could have developed in a strictly equivalent manner over long time periods. Sooner or later one must become superior to the other.

Until now I have presented natural selection and drift as effective mechanisms for the automatic elimination of certain competitors. But irreversibility does not rule alone in nature. There are also cycles and equilibria. If lions ate all the deer, they would die in their turn through lack of food. The existence of the strong depends on that of the weak. If there is a fluctuation of lions so that they are in excess at a given moment, the herds of deer will be decimated. Deprived of food, the lions will leave less descendants and in the following generation there will be a fluctuation of lions in the opposite direction. So there is no amplification of fluctuations. This type of relationship between species was studied in the early days of two disciplines: epidemiology (certain diseases manifest themselves in a cyclic manner in populations) and ecology, after the lead given by Lotka and Volterra. Population geneticists are now very interested in this approach. The school of Mayr, Lewontin and Maynard Smith refers to games theory in an attempt to inject some life into an evolutionism which until now has been too abstract and to discuss problems in a qualitatively varied way. An example will illustrate the difference between the old and new points of view. Most two-sexed animals and plants reproduce to give equal numbers of males and females. Why? The classical Darwinist will say that numerical equality between the sexes confers selective advantage because it maximizes the number of couples. But, with Fisher, we can pursue a completely different line of reasoning. Let us assume that the ratio of males to females (M/F) is genetically determined. In a population in which M/F is large (and which thus contains few females), a variant which tended to produce a high proportion of females in its offspring (low M/F) would be selected for, since its descendants would find partners easily. So when M/F is high in a population, individuals with descendants having a low M/F ratio are at an advantage. A high M/F ratio cannot therefore be maintained in a stable way in a population. Similarly, a low M/F ratio would not be, in the terms of Maynard Smith and Price, an 'evolutionarily stable strategy'. The only stable strategy, except in very special cases, is equality between males and females.

Although natural selection, by eliminating less-fit species, is on the whole a progressive factor, the selection which operates at the sexual level seems to work in the opposite direction. The shimmering plumage of the peacock which brings him success in finding a mate makes him vulnerable to predators. The antlers of deer, arms for dominating rivals in contests for females, hinder flight through the forest. Several independent lines of deer whose antlers did not stop growing have passed into oblivion. It is the female who, through her love of uselessly ornate heads, leads the species to its doom. But I suspect that indulgence and misogyny underlie this reasoning and it is not without pride that I cite the work of the Frenchwoman Claudine Petit, who performed experiments to reassert the value of female sexuality. Studying mating in fruit flies, she was able to show that females in the presence of males of different morphologies showed preference for the less common types. By doing this they help to maintain gene diversity in the population which, as we shall see later, is essential for the destiny of the species. This fundamental experiment of copulation genetics is a source of more general ideas on 'frequency dependent' selection. Certain characters are advantageous only when they are rare and for others the opposite applies.

In sexed populations, each individual possesses two sets of chromosomes, one from its father the other from its mother. If a gene is present in two allelic forms A and a in the population, there will be three categories of individual: A/A, A/a and a/a. For two genes A and B and their alleles a and b, there are ten possible genetic formulae: AB/Ab, AB/aB, etc. The number of combinations rapidly becomes enormous. With the exception of true twins, it is highly improbable that two human beings on this planet have the same genetic constitution. Any two men, taken at random, will differ by about ten per cent of their genes. In addition to genetic variability we must also consider the fact that a gene does not specify a precise character (such as a height of 1.76 m) but a gradation which we have to describe with a probability distribution and its statistical variance. The expression of a gene and its usefulness can depend on the presence of other genes, on external conditions, on the individual's past, etc. But most debates in population genetics consider simpler situations: one or two genes and their alleles, and no more. It is often suggested that genes have independent effects which are additive and that consequently a heterozygote A/a has properties intermediate between those of the homozygotes A/A and a/a. The reverse case is interesting. If A/a is more advantageous than A/A and a/a, the two characters A and a will be maintained in equilibrium in the population (selection favours A/a and therefore the simultaneous maintenance of both alleles). In West Africa individuals suffering from sickle-cell anaemia have a mutation in the gene for haemoglobin β. They are homozygous (a/a) for the mutated gene. But normal A/A individuals seem to be less fit than the heterozygotes A/a because they are less resistant to malaria; their molecular good health benefits the malarial parasite. Conversely, if the heterozygote A/a is less fit than the two homozygotes, we can predict accelerated fixation of one or other allele.

The greater the genetic variability in a population, the more chance it has when faced with important environmental changes of finding within itself combinations of genes which will allow it to adapt. Fisher, who founded this idea on statistical calculations, modestly gave it the name 'The Fundamental Theorem of Natural Selection'. The promoters of the 'green revolution' had without doubt lost sight of this idea. They developed high-yield varieties of wheat which they substituted for natural wheat; but disease attacked the new, genetically pure wheat and disaster resulted. Since all the plants were similar, they were all equally vulnerable. Since then, people have been frantically reconstituting the old strains. In a very heterogeneous population, the most advantageous variant is diluted. The discrepancy between the best performance of which a population should be capable in any given circumstance and the actual performance achieved is called the genetic load. It is the price of all-risks insurance for the population, since variability allows it to deal with the unexpected. But although variability 'dilutes' good genes, it also camouflages the bad ones and prevents their being selected against (of course the 'value' of a gene depends on the environmental circumstances). Mixing of populations increases variability. Human communities which used to live in relative isolation are re-sorting their genes. But the reverse phenomenon can occur – the founder effect. Whole regions can be populated by the descendants of a very small group of individuals. The resulting population is very homogeneous. Almost perfect isolation is required to maintain this homogeneity since, as Kimura has shown, it is only necessary for a few foreign individuals per generation to integrate into the population for its genes to be equilibrated with those of the foreign populations. If some ab/ab individuals are introduced into a homogeneous population AB/AB, the genes redistribute themselves, but nevertheless correlations between the presence of a and of b in individuals will still be found over several generations. This is 'linkage imbalance', which can indicate some of the ways in which a population may have arisen.

Mutations are a permanent source of novelty. Their incidence is genetically controlled. A gene has, on average, one chance in 100 000 of mutating from one generation to the next. From Fisher's point of view, this 'mutation pressure' on the population is beneficial to it. By increasing variability, it allows further gains in adaptability. On the other hand, it is true that unfavourable mutations are much more common than favourable ones. The population increases its genetic load, deviates from the optimum and perhaps is then doomed. If we cause a large number of mutations in a population, for example using ultraviolet radiation, and at the same time we subject it to selective tests, it will have more chance of extricating itself successfully than a non-irradiated population. But when a normal population and a fast-mutating one (in which the rate of spontaneous mutations is always higher than normal) are grown together in a stable and well-defined environment, the fast-mutating one does worse, as Ayala showed in populations of flies.

In calculations in population genetics, mutation is always treated as a corrective factor and never for what it is, an innovative factor. Evolutionists who believe their calculations almost reach the point of denying the existence of novelty. A favourable mutation for the mouse is one which helps it to run faster to escape from the cat. For the cat, a favourable mutation is one which gives it better legs to catch the mouse. Both cat and mouse evolve without qualitative changes, while keeping to their respective positions. Evolution is a race in which the competitors go as fast as possible just to stay in the same place and maintain their relative positions. This concept, supported by Van Valen, has gained many followers since economic inflation became established in the West.

With Mayr let us now look through the other end of the telescope. By emphasizing the enormous variety of possible combinations of genes I may have given the impression that living creatures belong to a sort of continuum in which properties vary imperceptibly from one individual to another throughout the animal kingdom. This is not the case. Living creatures can be classified into discrete groups. Alas, men and monkeys cannot cross-breed and exchange their genes and yet we do share common ancestors with them. Events occur which cause species to diversify and then separate into distinct species. Among these events are chromosomal accidents such as translocations, inversions, fusions or dissociations of chromosomes. A cross between two creatures whose chromosomal make-ups differ too greatly is sterile, as a result of difficulties in egg division. A major chromosomal change condemns the individual who carries it to sterility but a more moderate change leaves it with some chance of reproducing. Its descendants are more fertile when they cross-breed among themselves rather than with individuals having 'normal' chromosomal formulae. They thus end up by forming a nearly autonomous subpopulation. Other factors can contribute to separation of a species into autonomous subpopulations: geographical isolation, self-reproduction in plants and crosses between closely related individuals (inbreeding). Isolated subpopulations gradually diverge from one another. One reason for this is that mutations do not necessarily have the same selective advantage in two subpopulations. In addition, mutations which are almost neutral become fixed at random. So there is no reason why isolated subpopulations should accumulate exactly the same mutations.

In Man there are (or were) relatively isolated subpopulations, but the separation was not sufficient to lead to a split into separate species. We are now going in the opposite direction towards a mixture and so, paradoxically, towards diversification since new gene combinations will be possible. After isolation, a subpopulation will show a set of particular characteristics which are not necessarily dependent on one another. For example, a statistician would be able to establish 'statistical correlations' between skin colour and height among Brazilians: black-skinned men would be taller than men with white skins. The conclusion would be 'highly significant' if one trusted statistical methods. However, there is not the slightest causal relationship between being black and being tall. If the statistician had gone to a territory inhabited by pygmies, he would have understood this. It is important to emphasize this at a time of a racist offensive on a vast scale to 'prove', with the aid of genetic studies and statistical tests, a claimed superiority of men with white skins. The best statistics in the world do not permit observations to be raised to the status of permanent laws.

As early as 1930, Sewall Wright saw evolution as the result of two processes: fixation in subpopulations of mutations which when taken separately are slightly advantageous or disadvantageous (that is fixations in which random drift plays a role), and effective selection between the subpopulations which differ by several characters simultaneously. In fact, except for statisticians like Fisher, the whole is not only the sum of the parts. When two subpopulations have clearly diverged, they are far from being equivalent in the struggle for life even if each difference, taken in isolation, is selectively neutral. Contemporary evolutionists, led by Kimura, have brought gene fixation by drift into line with current taste; but the second point, which for Wright was the natural complement of the first, has been lost from view. How difficult it is to grasp both sides of an argument at once!

Diversification of species does not continue indefinitely. Species and whole classes become extinct. Why have the diplodocus, the ichthyosaurus and other innocent creatures of the Cretaceous period left no descendants? Why was there no mutation to put them on the road to salvation? If, as is said, they perished because of their gigantic sizes, why could they not have mutated towards smaller sizes? These questions are pure delight to the anti-Darwinists. Recognizing that several independent lines were all able to evolve in the same direction, and all to their doom, we must assume like Haldane, in the memorable lecture which he gave at Prifisgol Cymru in Aberystwyth, that 'the evolutionary process somehow acquired a momentum which took it past the point at which it would have ceased on a basis of utility'. In the case of the dinosaurs, we may imagine that an intense selection acted at an early phase of development, favouring large newborn. But a developmental programme which is advantageous at birth can have awkward consequences at reproductive age. Again, the use by a given species of a certain evolutionary strategy (mutation rate, mode of crossing) may pay off in a whole series of situations but finish by stumbling over a special case which was not 'foreseen'.

Another phenomenon which has posed problems is that certain characters (notably size) evolve slowly and monotonously along certain lines. If it was advantageous for a horse to be large, why did large mutants not rapidly replace their smaller fellows? One answer is that an important change of size can be a source of difficulties: embryonic development poorly adapted to the size of the womb, the need for a more powerful heart, etc. The overall result for the individual is apt to be only slightly advantageous or negative. A smaller size increase, more compatible with the individual's other characters, would be more advantageous. Besides, there is nothing to say that changes in size are necessarily genetic in origin. Gene expression can be affected by changes in the environment: climatic conditions, food, etc. Since Darwinism copes well with the gradual evolution of characters, the objection was raised that: there seem to be some gaps in relationships among fossil species. Species U should logically descend from species V. Now U and V are sufficiently different for there to be good reason to postulate the existence of an intermediate which the palaeontologists do not find. We can answer this too. We may suppose that evolution is made up of calm periods in which characters change slowly and progressively and brief, agitated periods in which events occur rapidly. Since these transition periods are short, they leave few fossils, hence the dificulty of finding missing links.

The theory of evolution by natural selection thus has an answer for everything. It rests on a type of reasoning which is so evident that it cannot be falsified. In that case it is not a scientific theory, the followers of Popperian obscurantism will say; a point on which, with Lewontin, I find them wrong. For every scientific theory rests, in the last analysis, on very general principles which are not or cannot be subject to experimental test. Has it ever been proved by experiment that the Earth revolves about the Sun? Like the system of Copernicus, Darwinism is a general conception about which a large number of facts can be ordered. But its heuristic value remains low. One cannot avoid making a detailed analysis of the concrete situation in each particular case. And population genetics is far from having exhausted the study of all the simplest test systems, so evidence of new effects may still be gained. For example, serious attention has never been paid to the possibility that a gene might exert its effect only in the second or third generation. Reasonable molecular scenarios can be invented for this. Certain zones of the chromosome, the 'hot spots', mutate very easily. I do not rule out the possibility that in certain cases there is a special mechanism allowing changes of genetic programme to be made in a determined direction, in response to a determined change in external conditions. An individual would potentially contain several programmes, each one mutable into any other. This is a suggestion from molecular biology for population genetics.

According to the ultra-selectionist view, in which each mutant is rapidly fixed or eliminated, populations must be very homogeneous. For the neutralists, on the other hand, each gene drifts slowly and since new variants constantly appear by mutation, populations must be very heterogeneous. To settle this, one measures molecular heterogeneity by systematically comparing the proteins of individuals in a population. One extracts a particular protein, haemoglobin for example, and measures its electrical charge (roughly speaking, the difference between the numbers of positively and negatively charged amino acids). Among the variants of a protein some have the same charge while others (about a third) have a different charge. If in 1000 individuals, 988 have a haemoglobin of the same charge, and twelve one of different charge, it is reasonable to conclude that in this sample there are in fact about thirty individuals with abnormal haemoglobins. One thus measures the 'heterozygosity level' of the population for the protein concerned – a sort of mean quadratic difference. The 'heterozygosity levels' for many enzymes have been measured during the past few years, since Lewontin and Hubby introduced the technique of electrophoretic detection in 1966. The mean level for regulatory proteins in man is 0.13 and the heterozygosity for other proteins 0.05. What will these results prove in the last analysis? Each new result on heterogeneity is dissected and used by the two camps: the neutralists who have ended up giving some place to selection and the hard-line selectionists who are beginning to think that certain proteins are less subject than others to selective pressures. We await with bated breath the outcome of the continuing battle of the giants, between the supporters of the half-empty bottle and those who claim that it is half-full. Molecular biology has some original items to add to this file which are not generally known to population geneticists. They will be presented in a completely new light from Chapter 12 onwards.

Recently some physicists attempted a take-over bid of the study of evolution, assuming that, with their superior science of differential equations, they would be able to reveal the essential truth of the phenomena. What came out of this was the application of mathematical treatments to the simplest test case of population genetics: bacterial competition in the chemostat. Or rather, having changed the words, these physicists described competition between DNA molecules in the prebiotic soup, attributing to the molecules the same reproductive properties as to bacteria. Having resolved the simplest problem, 40 years after the evolutionists, the physicist is confident that he has at last given Darwinism a scientific foundation. In fact the only point worth discussing in these treatments is: is it right to transpose the classical treatments of population genetics to molecules? I answer without hesitation: no, and I will justify this position in the next chapter. In the final chapter I will outline the kind of mathematical treatment which adheres to the set of problems gained in studying molecular evolution. Neither Darwinist nor anti-Darwinist, they seek to answer questions which are outside the problems borne by the Darwinists.

When the biologist is asked why the cell does something in such a way, he is never at a loss: in the past this way had a selective advantage over the others! Too many theoretical treatments of molecular evolution and too many discussions of the origins of the genetic code have over-generous recourse to 'selective advantages', which are used to fill in every hole in the argument. By revealing the complexity of population genetics I wanted to arm the reader for critical appreciation of other works or help him in arguments which he would be involved in. We should beware of terms such as selection and selective mechanisms in biology, since this language is often the camouflage of ignorance.

8 Prebiotic replication

Does life start with DNA? Many theories of evolution suppose that DNA molecules were able to replicate on their own in the prebiotic soup, without enzymes, gradually influencing and mastering their environment. The code would then have been merely the means for a DNA molecule to replicate even faster by using a coded DNA polymerase. Here we shall examine the first floor of this house of cards. Can DNA really replicate without an enzyme? Twelve years of experiments, mainly in Orgel's laboratory, allow us to define what prebiotic replication of nucleic acids might have been like.

The first two successful experiments revealing the outlines of replication were described in 1966 by Naylor and Gilham, and by Sulston and co-workers in 1968. The first two authors were not interested in the origin of life but hoped to find a general method for synthesizing nucleic-acid sequences at will. They mixed a DNA chain containing only A residues with oligomers of T:

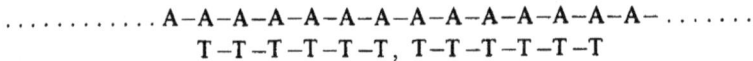

$$\ldots\ldots\ldots\ldots A-A-A-A-A-A-A-A-A-A-A-A-A-A-\ldots\ldots\ldots$$
$$T-T-T-T-T-T, \; T-T-T-T-T-T$$

In water, near 0°C, complementary associations occur. To join the two thymine hexamers, a water molecule has to be eliminated. Naylor and Gilham used a dehydrating agent, carbodiimide. After incubation for four days at −3°C, five per cent of the thymine hexamers had formed dodecamers. Repeating the same experiment with pentamers, they obtained three per cent decamers. It is paradoxical that molecules can be dehydrated in an aqueous medium. Carbodiimide is avid for free water, but it reacts even more vigorously with the water potentially present in certain compounds.

Sulton's success in obtaining a rough sketch of replication came through proceeding unit by unit. A complex such as

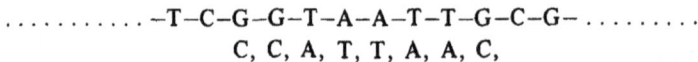

$$\ldots\ldots\ldots\ldots -T-C-G-G-T-A-A-T-T-G-C-G-\ldots\ldots\ldots$$
$$C, \, C, \, A, \, T, \, T, \, A, \, A, \, C,$$

in which a DNA chain attracts complementary monomers which obediently form pairs is Utopian, since the affinity of bases for their opposite numbers is too low. However, the association becomes possible when several interactions are combined, as with the thymine oligomers. Even in this case, the temperature has to be reduced to 0°C for stable complexes to be formed.

However, for monomers, there are two favourable cases. Monomers of A can associate with polymers of U, binding to two chains at the same time so that a triple helix is formed:

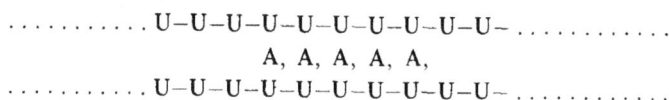

```
.......... U–U–U–U–U–U–U–U–U– ...........
             A, A, A, A, A,
.......... U–U–U–U–U–U–U–U–U– ...........
```

Using carbodiimide, Sulston and colleagues succeeded in condensing monomers of AMP and obtained forty per cent dimers after 17 days and even six per cent trimers. Fine analysis of the products formed was disappointing. Instead of obtaining the 3'–5' bonds present in biological RNA, they found a majority (sixty per cent) of 5'–5' bonds which, in other words, linked the two phosphates of the AMP molecules. This method of linkage does not allow regular extension of the chain. More than four-fifths of the remaining forty per cent of the linkages were of the type 2'–5', different from but closer to the biological type (Fig. 14(b)). Similar results were obtained with another special complex in double-helical form:

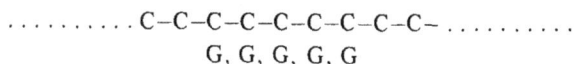

```
.......... C–C–C–C–C–C–C–C–C– ..........
             G, G, G, G, G
```

Success depends on the fact that monomers of G tend naturally to stack on each other, even in the absence of a complementary template, and behave as if they formed true oligomers.

No effort was spared to improve the efficiency of oligomer formation from monomers on using complementary templates. The use of carbodiimide was abandoned and instead monomers were chemically activated by attachment of imidazole groups. In this way the 5'–5' linkages were eliminated but not all the 2'–5' bonds could be got rid of. Moreover, everything depended on having the two types of complexes shown above. Copying any chain, residue by residue without enzymes, hardly seems possible. There remain the methods using oligomers.

Shabarova and Prokofiev had the idea of activating the terminal phosphate of an oligomer of A with a derivative of phenylalanine. No doubt they secretly hoped that copying of the U residues would be facilitated by the presence of the amino acid which has UUU as its codon. Usher and McHale also condensed oligomers of A in which the terminal phosphate had been made cyclic. The product formed contained ninety-five per cent 2'–5' linkages and five per cent 3'–5' bonds. But they point out that the latter were much more resistant to hydrolysis when they formed part of a double-chain complex than were the 2'–5' bonds. The rates of hydrolysis are in the ratio of 1 : 900 at 40°C. So the 3'–5' bonds, which are in the minority at the beginning, survive longer. Shamovski carried out a very careful physico-chemical study of complexes between polymers of T and oligomers of A. He concluded that the oligo As are arranged head to tail in the triple helix:

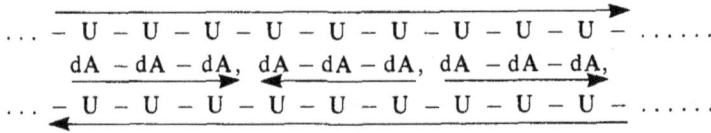

$$\begin{array}{c}
\overrightarrow{}\\
\ldots - U - U - U - U - U - U - U - U - U - \ldots\ldots\\
dA - dA - dA, \quad dA - dA - dA, \quad dA - dA - dA,\\
\overrightarrow{} \qquad \overrightarrow{}\\
\ldots - U - U - U - U - U - U - U - U - U - \ldots\ldots\\
\overleftarrow{}
\end{array}$$

For my part, working in Orgel's laboratory, I formed various chains with oligomers activated with imidazole. It appears that condensation on a matrix is generally easier to achieve block by block rather than unit by unit. The proportion of 3′–5′ linkages is extremely variable from one situation to another. The two complexes below:

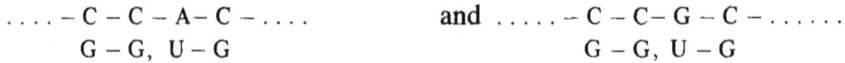

$$\ldots - C - C - A - C - \ldots \qquad \text{and} \qquad \ldots - C - C - G - C - \ldots$$
$$\quad G - G, \ U - G \qquad\qquad\qquad\qquad G - G, \ U - G$$

are equally favourable for obtaining condensations; the non-complementary pairing G.U is here almost as effective as A.U.

The enzymes which join oligomers, DNA or RNA ligases, might in fact have preceded DNA polymerases during evolution. They have a curious way of performing their reaction. Let us suppose that it is necessary to join the two oligomers pGpCpGpC and pCpCpTpA on a complementary matrix. The DNA ligase uses an ATP molecule to activate one of the oligonucleotides, forming a small appendage which is excised later while the other oligonucleotide is spliced into the nick:

$$\text{pGpCpGpC} \overset{\frown}{\underset{\text{Ap}}{}} \text{pCpCpTpA}$$

Such a mechanism, far from being aberrant, uses the kinds of compounds which one would expect to find in prebiotic conditions. Oligomers of the type XppYp. . ., which therefore contain a 5′–5′ bond between X and Y, are obtained in preference to 'regular' oligomers, provided that one doesn't try to deliberately avoid them.

Condensation reactions on a matrix are only one stage in the long reaction chain of which an authentic prebiotic replication would be comprised. There are still formidable difficulties to be overcome at every other stage.

Although adenine is easily formed by prebiotic pathways (from hydrogen cyanide) and guanine is obtained without too much difficulty, the synthesis of pyrimidines, particularly that of cytosine, remains a mystery. Sugars are formed easily from formaldehyde. As well as ribose (which forms a few per cent of the total) various other sugars are obtained: fructose, cellobiose, xylulose, etc. If one tries to link ribose to adenine by non-enzymatic means, one obtains several compounds which cannot integrate into double helices; biological-type adenosine represents only four per cent of the products formed (Fig. 14(a)). If we want to attach a phosphate onto an adenosine we preferentially obtain AMP with a phosphate at the 2′ or 3′ positions or again with a 2′–3′ cyclic phosphate. But for enzymatic replication, it is the AMP with a phosphate at the 5′ position which is interesting. And this is just the beginning. Having obtained (after a great deal of trouble) matrices and oligomers, we can join the latter and form chains complementary to the matrices. To have true replication of the initial chain, we should separate the

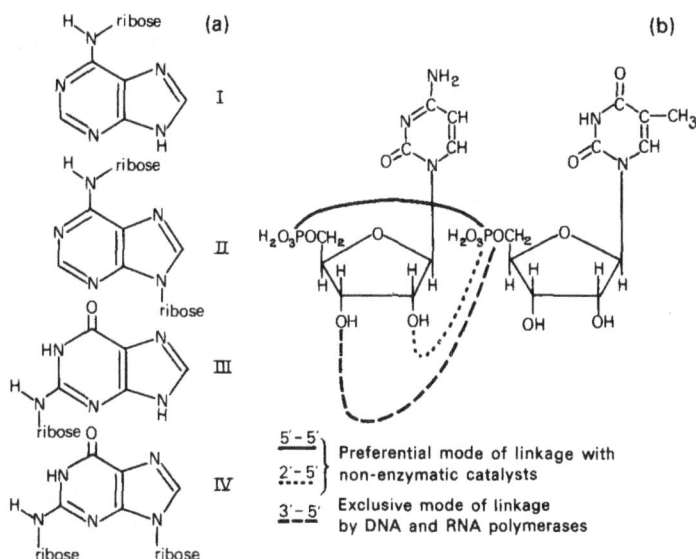

Figure 14 Pathways of prebiotic chemistry. When a mixture of ribose and adenine, adenosine, guanine or guanosine is heated, it forms compounds preferentially of types I, II, III and IV respectively (a). None of these can be used in RNAs and no prebiotic mechanism is known which would lead to compounds of the biological type. Also, experiments on condensation on matrices in the absence of enzymes usually result in bonds between a 5′ carbon and another 5′ carbon or a 2′ carbon, whereas the cellular mode of polymerization links a 5′ carbon to a 3′ carbon.

two complementary strands and repeat with the new one what was done with the old. Cycles should be introduced into this set of operations, with, for example, a daytime phase during which the chains separate because it is hot and a nocturnal phase of association and polymerization. But when the chains are separated to be used as separate matrices, how do we prevent their reassociation in pairs as before? Partial answers can be found to each of these difficulties. But I would not hold out much hope of success for the chemical engineer who wants to construct a 'nucleic-acid replication machine' which functions without enzymes.

It is worth while reflecting on the possibility of the existence of replicating systems without nucleic acids. In one sense, a crystal which grows, reproducing an atomic arrangement indefinitely, then breaks and forms two crystals which grow in their turn is an object which replicates. Better still, a crystal may contain imperfections which are also propagated. Cairns-Smith has studied these possibilities closely and considers that certain clays may have properties of a primitive replicating system. These clays have a stacked sheet structure; the imperfections of one sheet are transmitted to the next. Since clays also provide an ideal environment for prebiotic reactions, Cairns-Smith does not have too much difficulty in constructing a theory in which life begins with clays which grow, reproduce and influence their environment. Staying closer to present-day life, we may imagine polymers chemically simpler than DNA or RNA but nevertheless capable of forming complementary pairings. Thus the Pithas, using as unbiological a matrix as poly (1-vinyl-uracil), which is a linear polymer of U residues but without ribose or phosphate, were able to condense AMP with a higher efficiency than in the reaction with classical poly-U.

But polyvinyl-uracil has a staggered conformation and cannot be part of a double-helix. Marlière constructed models and saw that some simple polymers to which nucleotide bases were hooked, could well form good helices when the repeating backbone unit contained an even number of atom links. Of these, some polymers of modified β-amino acids like Poly A (adenyl-ethyl) β-alanine were synthesized by Takemoto, but not tested as templates.

Lohrmann, Orgel and Inoue try to replace standard polunucleotides, with monomers closely related to the biological ones. The miracle compound today is a nucleoside monophosphate, whose 5′ phosphate is activated by a 2-methylimidazole group. Activated monomers of G can be joined efficiently and accurately on a Poly(C) template. The links are predominantly 3′-5′. When the reaction is carried out in a triple-stranded geometry, the products start with a 5′-5′ linkage, thus: GppGGGGGGGGG . . . . , much like the natural messenger RNAs synthesized by RNA polymerases in eukaryotes. This feature can be explained if triple-stranded structures favour, as shown on p. 60, head to tail arrangements of monomers and oligomers of G.

I see prebiotic replication as the result of several activities, according to the scheme in Fig. 13 for the formation of palindromes. At first, nucleic-acid chains of any sequence whatever fold up or associate in twos or more, matching up any complementary sequences as best they can. Afterwards, the badly paired segments are excised and finally the holes are refilled. Thus any sequence evolves towards a complementary double chain. The chains separate, fold up on themselves and the process starts again. If we start with random sequences and subject them to activities which cut them and rejoin the pieces we should, under certain constraints, observe a tendency to form long chains in double helical form. By combining the same activities but with different constraints, we could form different structures. Will we understand why tRNA is a cloverleaf by following this approach? The experiment is yet to be done.

9 The genetic code

It is thought that Dounce was the first person to formulate the central concept in molecular biology: the existence of a code of correspondence between nucleic acids and proteins. In his 1952 paper, Dounce proposed the following: genetic information contained in a DNA sequence is first transcribed into RNA; each set of three nucleotides (now called codons) corresponds to an amino acid; the sequence of codons in a gene determines the order of amino acids in the protein which is the product of the gene; the correspondence is established in an indirect manner with adaptor proteins which recognize both the codon on the one hand and the amino acid on the other. We now know that there is a supplementary link between the codon and the amino acid which Crick postulated: transfer RNA. The adaptor protein which, through habit and for linguistic convenience, I shall call the activating enzyme, links the amino acid to the transfer RNA and the latter recognizes the codon (*see* Chapter 2). On many points, the dominant concepts in the period immediately after the discovery of the structure of DNA represented a clear retreat from Dounce's proposals. Information scientists, and not only they, invented all sorts of exotic codes which were refuted by Brenner.

Dounce is rarely cited. From 1964 to 1969, at the height of the period when the genetic code was being worked out, his article was mentioned only twice in scientific journals. Was he too modest to draw attention to his work and establish the authorship of his ideas? Presumably, like so many others, he was overshadowed by the domineering Cambridge school – now in decline – which imposed its concepts on the scientific press and ignored everything not proceeding from itself or that had not received its prior approval.

The genetic code is now entirely worked out (*See* Table 1) owing to the work of hundreds of researchers, especially that of the teams led by Nirenberg, Khorana, Brenner and Yanofsky. The set of sixty-four correspondences between codons and amino acids can be represented pictorially and several artists have attempted to give this keystone of molecular biology a represention which strikes the imagination. The most commendable one is the circular and radial representation by Ratner, master of Siberian cybernetics. But a square table has come into general use, which exhales a scent of hidden correspondences when the four bases are arranged in the order U, C, A, G. The table of the code (page 15) reveals many regular features. As foreseen by Eck, Woese and Sonneborn, the code is subdivided into families of codons; it is systematically degenerate. Two codons XYU and XYC, which start in the same way but have different bases (U or C) at the third position, still code for the same amino acid. Two codons XYA and XYG almost always code for the same amino acid. Let us look at the table of the code column by column. Amino acids in the same column are generally of similar polarity. Thus all amino acids having a U in the second position of their codons (those in the first column) are hyrophobic, and amino acids with an A in the middle position of the codons are rather hydrophilic*. Another regular feature is that 'small' amino acids like glycine have highly degenerate codons (each of these amino acids occupies a whole box in the table). In contrast, large amino acids like tryptophan or methionine are coded for by one or two codons – with the exception of arginine, which has six. The position of the amino acids in the table of the code is well correlated with their taste, sweet or sour. More generally, as observed by Mitsuyama, *yin* amino acids like threonine and serine correspond to *yang* codons and vice-versa. Lastly, amino acids which are closely related chemically or which are produced along the same biosynthetic pathways are also close neighbours in the table: each codon needs only one base to mutate in order to be transformed into the codon of a similar amino acid – but there are exceptions.

For all codons of the type GGX, GCX, CGX and CCX, degeneracy is complete in the third position. Whatever X is (U, C, A or G), the same amino acid is coded for. Conversely, degeneracy is always incomplete for the codons AAX, AUX, UAX and UUX: these codons' boxes always contain two amino acids. Lastly, observes Orgel, when the first two bases are mixed (an A or U with a G or C), degeneracy is complete if the pyrimidine (U or C) is in position 2; if not it is incomplete. Let us assume for the moment that the association between the codon and the tRNA sequence which recognizes it (the anticodon) occurs as in DNA by complementary pairing, at least for the first two positions. Let us also assume that a G.C pair is much more energetic than an A.U pair. Then, following Crick, we can interpret the preceding rules in terms of the stability of codon–anticodon associations. If the first two bases of the codon are Gs or Cs, they will form two G.C pairs which are sufficiently stable on their own so that there is no need to 'specify' (?) the base in the third position. If the first two positions of the codon contain As or Us, they will form two A.U pairs with the anticodon. In this case another pair has to be added to make the association stable enough, so the base in the third position must be specified. The logic of this argument may perhaps not be irreproachable, but nevertheless I agree with Crick in reckoning that this type of regularity in the genetic code must be related to the different stabilities of G.C and A.U pairs. The criterion of stability which may have been decisive at the time of the origin of the genetic code should be qualified. The codon's affinity for the anticodon can be modulated by a whole series of factors, notably the presence of chemical groups attached to the anticodon or neighbouring bases. Modulation of affinities now acts to restrict the range of stabilities, a G.C pair being roughly equivalent to an A.U pair in the codon–anticodon association. This remark is based on indirect evidence: the very systematic measurements of Grosjean, de Hénau and Crothers on self-association of anticodons. Their results agree well with a whole series of more direct but disparate observations.

Having taken note of the degeneracy of the code, let us look in detail at the laws which govern codon–anticodon association. The first ideas on the question emanated from Eck and Woese well before the code was worked out. Taking as a starting point the hypothesis that all or nearly all codons are used, it might have been imagined that they would have been allocated at random: for example valine would be coded for by GUC and AGU, histidine by GUG, CAA and UAU, etc. Eck imagined that codons might be grouped according to systematic laws. For example, XUY and XCY would always code for the same amino acid; similarly XAY and XGY would always go together. When Eck developed his hypothesis there was nothing to favour the third position of the codon–anticodon association and he arbitrarily placed systematic degeneracy in position 2. Barring this, his hypothesis has been strikingly confirmed. But there is better still. Eck proposed that a single anticodon would suffice for reading each pair of synonymous codons. According to him, pairing would occur according to strict rules of complementarity in the non-degenerate positions (by G.C or A.U exclusively) and would be more flexible in the degenerate position (which we now know to be position 3). Thus we would have:

(Eck's rules)   G (anticodon) pairs with C or U (codon)
U (anticodon) pairs with A or G (codon)

The notion that a transfer RNA can read more than one codon was quickly confirmed experimentally. Phenylalanine has two codons but there is only one tRNA in *E. coli* for reading them. Although alanine has four codons, we find only two alanine-specific tRNAs in yeast. The major one was the first tRNA whose sequence was known. The triplet IGC was found towards the middle. Inosine (I) is a base related to adenine, often present in tRNAs but never in messenger RNA or DNA. Holley, who sequenced alanine tRNA, suggested that its anticodon was IGC and that inosine was able to pair indiscriminately with either U or C. Going further, Crick proposed that inosine also paired with A. So, for position 3 of the codon–anticodon association we would have:

(Crick's rule) I (anticodon) pairs with U, C or A (codon)

Let us recall that by convention, position 3 of the codon–anticodon association comprises pairing between position 3 of the codon and position 1 of the anticodon.

Combining Crick's rule with Eck's two rules gives us the celebrated wobble (or fuzzy recognition) hypothesis of Crick, which had a profound influence on all the work which followed.

The possibility of pairing G with U had previously been thought of by biologists, but only when it included tautomers. It was believed that a G.U pair existed only during the brief instant when G ceased to resemble itself and adopted the deceptive appearance of an A. There is nothing like this in the wobble hypothesis; bases are always in their normal forms. Unorthodox pairings then have different geometries from the canonical G.C and A.U pairs. Crick studied on paper all possible base pairings and retained those which seemed plausible. Besides the canonical pairs and the non-classical pairs invoked by Eck and Holley, he found the pairs U.U, U.C and I.A. He then showed that if U.U or U.C pairs were used systematically in reading at position 3 of codons, absurdities would result. So it was neccessary to limit pairings to those mentioned previously: G.C, A.U, U.G (or G.U), I.C, I.U, I.A. These pairs have a geometry which is quite closely related to that of the canonical pairs.

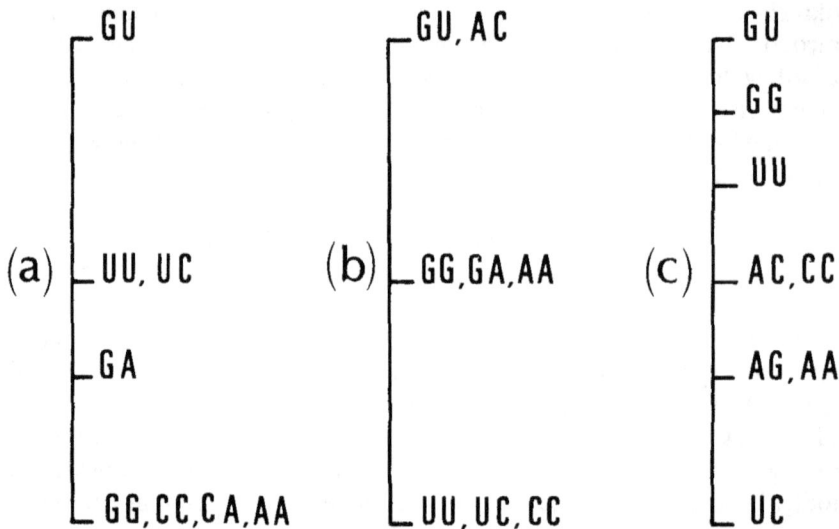

(a)	(b)	(c)
⌐ GU	⌐ GU, AC	⌐ GU
		⌐ GG
		⌐ UU
⌐ UU, UC	⌐ GG,GA,AA	⌐ AC,CC
⌐ GA		⌐ AG, AA
⌐ GG,CC,CA,AA	⌐ UU,UC,CC	⌐ UC

Figure 15   Three views of non-complementary base-pairs. (a) According to the wobble hypothesis (93) GU base-pairs are easily accommodated in the third position of the codon–anticodon association. UU, UC and GA are feasible but require a severe distortion of the sugar-phosphate backbones. The other pairs are ruled out. (b) Topal and Fresco (253) consider that most replication and translation errors are due to tautomeries, as in the case of GU and AC, or to flipping over of purines, giving rise to GG, GA and AA errors. The pyrimidine–pyrimidine pairs are forbidden. The ladder in (c) is based upon the consideration of double-helical sections in known nucleic acid structures. By assigning to every base-pair an energy value according to the ranking given in (c), one predicts correctly the secondary structures in both the tRNA and the 5S RNA molecules (200, and work in progress).

Hence Crick's idea: in the first two positions of the codon–anticodon association there is strictly complementary base pairing with rigidly defined geometry, while in the third position a little wobble or blurring would be possible so that pairs which were not too different in their geometry from the canonical pairs could be formed.

For about 10 years, nearly all experimental observations were declared to be in agreement with the wobble hypothesis, even though certain obvious incompatibilities had been detected early on. For example, certain bases derived from U, which I shall call V and 2S–U, were found in anticodons and formed unexpected pairings with the third base of codons. The base V is very widespread in bacterial anticodons. It has been studied particularly well by the Japanese scientist Nishimura whose rule states:

(Nishimura's rule) V (anticodon) pairs with U, A or G (codon)

This type of degeneracy must have been observed quite early but was swept under the carpet because of its non-conformity with the prevailing dogma. Nishimura had the courage to say loudly what he had observed. Nishimura's rule, which is just as soundly established as Crick's, does not always have the right to be cited in molecular biology texts as official as Watson's *Molecular Biology of the Gene*. Even for American students, it is not good to tell the whole truth. As for the base 2S–U, it is generally given the following property:

(Ohashi's rule) 2S–U (anticodon) pairs with A (codon) exclusively

Several groups have attempted to synthesize proteins in the test tube and to determine the tRNAs capable of reading each codon. The results converge and indicate that the reading laws are much more flexible than those predicted by wobble (Fig. 15).

In fact, the search for detailed laws concerning the reading of codons seems futile to me. The approach which consists of cutting the codon–anticodon association sandwich into slices, with certain rules of pairing for the first two slices and others for the third, leads to a dead-end and has now largely been overtaken. I had the good fortune, when starting in molecular biology, to glimpse a possibility on a completely different level from that of the classical explanations. It seemed so preposterous to those around me that I had to wait 4 years before being able to publish it, and then thanks to the recommendations of François Gros and Francis Crick. This is the missing-triplet hypothesis, which paved the way for the most recent work on the precision of protein synthesis. Before explaining it, I shall introduce some later developments. First of all, we must rid ourselves of the notion of the genetic code as a static table or correspondence between two vocabularies and examine closely the process by which proteins are synthesized.

Let us pass over the first phase of the operations: initiation, in which the codon at which translation begins is chosen using a complex procedure which is still not entirely worked out. Once the first codon has been chosen, the messenger RNA is translated codon by codon during a set of operations which constitutes the 'elongation cycle'. A cycle starts, after incorporation of the nth amino acid into the peptide chain, when the $(n + 1)$th codon is positioned ready to be read on the ribosome. About forty tRNA species circulate around the ribosome, some carrying the amino acid which corresponds to the codon. An association is formed between a codon and a tRNA which has a lifetime of variable length. While the tRNA is bound, precariously, to the ribosome, the ribosomal proteins attempt to catalyse the formation of a peptide bond between the amino acid carried by the tRNA and the protein chain which is being formed. If the codon–anticodon association is highly unstable, its duration being particularly short, there is a strong chance that the tRNA will have left the ribosome before the peptide bond can be formed.

On the other hand, if the association is highly stable, the ribosome will have ample time to catalyse the transfer of the amino acid to the peptide chain. We are no longer speaking of laws of pairing but of durations of associations. A brief association is hardly effective, whereas a longer-lasting one ensures incorporation of the amino acid. The ribosomal cycle used to be divided into stages. First the tRNA was chosen and then the peptide bond was formed. In the new description, the choice of the tRNA and formation of the peptide bond are not dissociable. The tRNA can leave the ribosome at any moment, notably during attempts to form the peptide linkage. The mechanism of choice between what could be called correct associations (stable) and incorrect ones (brief) lies in the race between two possible and independent events, disruption of the codon–anticodon association and formation of the peptide bond. This conception has been amply confirmed by experiment, notably through the work of Spirin's school.

Let us call the average duration of the association between a codon and an anticodon t, and the probability that the amino acid is incorporated into the chain after the association has been formed $p(t)$. What does the function $p(t)$ resemble? We can make some reasonable remarks. Let us suppose that for an association of 10 ms the probability of incorporation is close to one (0.99, for example). With a duration 10, 100 or 1000 times longer, the probability of incorporation will be more or less the same since there is not much margin between 0.99 and the maximum, 1.00. So beyond a certain duration, all associations are equally effective. Things are different for short-lived associations. If there is one chance in 1000 of forming a peptide bond in 1 ms, then in 2 ms the chances are likely to be twice as high. If we compare the durations of a correct association and an 'undesirable' one for the same codon and find that they are in a ratio of 100, the corresponding probabilities of incorporation of the amino acid will be in a ratio of 100 or less, and this is so for most known reaction mechanisms. But, in theory at least, there are particularly discerning mechanisms which, while using lifetimes in a ratio of 100 : 1, can lead to probabilities in a clearly higher ratio so that they amplify discrimination between correct and incorrect associations. The discovery of these mechanisms was the fruit of discussions which I had with Orgel when I was working in his laboratory. But the idea was already imminent, since it was simultaneously proposed by Hopfield, and Blomberg in Sweden also came very close to it. Mechanisms for increasing discrimination consume energy to obtain higher specificity. Quite apart from these mechanisms, fine analysis shows that there is, in any case, an energetic cost specificity. But this is difficult to evaluate. Hydrolysis of an ATP molecule would seem to be more than sufficient to form a peptide bond between two amino acids. In practice, however, the cellular machinery consumes at least three molecules of ATP or GTP to make a peptide bond. This excessive energy expenditure may represent the price of specificity. But, since the cost of making an amino acid is about ten times greater (about 12 ATPs for glycine, 78 ATPs for tryptophan), energy wastage during peptide-bond formation is moderate. Having said this, when the cell makes an enzyme 1100 amino acids long like β-galactosidase, when a protein 150 amino acids long should have been able to do the job, we may conclude that economizing energy is not the cell's major objective. Going beyond this we could argue that perhaps the 1100 amino acids of β-galactosidase allow it to be better regulated and intervene more specifically so that it saves energy.

Let us return to codon–anticodon association:

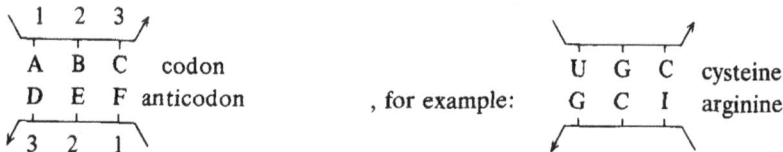

```
  \  1   2   3  /                              \             /
     A   B   C   codon                          U   G   C   cysteine
     D   E   F   anticodon     , for example:    G   C   I   arginine
  /  3   2   1  \                              /             \
```

It obeys the laws of nucleic-acid physico-chemistry. To evaluate its stability, it is necessary to take into account not only individual pairs A.D, B.E, C.F but also cross-forces between A and E, C and D, etc. It follows that certain non-complementary associations can be as stable as complementary ones: the fact that the pair C.F, for example, is highly unstable is compensated for by a very favourable diagonal interaction between B and F.

Let us imagine that an arginine tRNA with the hypothetical anticodon GCG could associate in a relatively stable manner with the cysteine codon UGC. How can errors in translating the codon UGC be avoided? One way would be to use for arginine a modified anticodon which associated less well with the codons. If the anticodon GCG is replaced by ICG, the anticodon's capacity to associate is reduced, since the I.C pair which is formed is less stable than the previous G.C pair. In this way both correct associations (with the codon CGC) and incorrect ones (with the cysteine codon UGC) are weakened. Then, when the arginine tRNA associates with the cysteine codon, the probability of forming a peptide bond is reduced, for example, from 0.05 to 0.005 and the rate of errors decreases. But the arginine tRNA continues to read its complementary codon roughly normally, the probability passing, for example, from 0.99 to 0.89. By weakening codon–anticodon associations, reading is made more precise, which is contrary to the intuition of most biochemists. According to this concept of codon–anticodon recognition, the principal role of inosine would not be to allow blurred reading at the third position, but to weaken the anticodon's capacity to associate, thereby avoiding errors. In sum, it is important for the cell to avoid using too strong an anticodon such as GCG: the triplet GCG cannot be used as an anticodon and is absent. That is why I called this idea the missing-triplet hypothesis. Ten years after its conception and six after its publication it is beginning to make its way.

This hypothesis went against the dominant concepts in two ways. First, it considered that non-complementary pairs, G.U for example, were quite 'normal' and so could in principle occur at all three sites of the codon–anticodon association. In the wobble hypothesis, on the other hand, something special was invoked in position 3 so that G.U pairs might occur. All recent results in nucleic acid physico-chemistry indicate that the G.U pair is quite normal, almost as stable as an A.U pair. After the publication of the wobble hypothesis, molecular biologists took about 7 or 8 years to recognize that the G.U pair is a normal component in nucleic–acid associations. More seriously, the missing-triplet hypothesis knocked flying the dogma that every biological explanation had to be structural, the myth of the structure–function relationship. According to the hypothesis, if I is present at position 3 it is not because associations in position 3 are particularly ambiguous but to avoid ambiguities *elsewhere*, in positions 1 or 2. Suggesting this was almost like raising the spectre of unemployment for the crystallographers who believed that the ultimate explanation of phenomena lay in determining the positions of atoms in molecules.

During the elongation cycle (Fig. 3, page 14) two tRNAs are attached to the ribosome, one carrying the forming protein, the other the last amino acid. The elongating chain detaches from the first tRNA 'looks for' the second and attaches itself to the amino acid carried by this tRNA. The first tRNA leaves the ribosome and the second takes its place while the messenger RNA advances so that a new codon occupies the site which has become vacant. Let us count the components implicated in genetic translation. There are twenty activating enzymes (one per amino acid), about forty species of tRNA, about fifty ribosomal proteins, three ribosomal RNAs 120, 1500 and 2900 nucleotides long (figures given for *E. coli*) and ten or twenty 'factors' – proteins which participate in initiation, elongation and termination of chains; plus a battery of enzymes in charge of preparing and maintaining the tRNAs (these chemically modify certain bases, repair their ends or degrade them when they appear abnormal). In all, thirty per cent of cell proteins serve only to carry out protein synthesis, which is comparatively slow. An *E. coli* ribosome only incorporates ten amino acids every second and eukaryotic ribosomes are even slower.

In principle every chemical reaction is reversible. Is it possible to reverse translation? Many authors have raised the possibility of reverse translation in which the protein would direct the synthesis of an RNA coding for it. If one extends reversibility to the whole chain of actions, genes → proteins → surroundings, one finishes by proposing a direct action of the external environment on the genome which, say these authors, would allow more rapid evolution than that caused by natural selection. In my opinion, those who propose such adaptive evolution are wide of the mark. For one thing, the reverse chemical reaction of genetic translation is not synthesis of RNA coded for by a protein but a much less interesting reaction: a backward movement of the messenger RNA coupled to hydrolysis of the protein, amino acid by amino acid. To translate protein into RNA would require a translation machinery completely different from the usual one. A new difficulty is encountered in going backwards and forwards between genes and proteins, that of avoiding ambiguities. Everyone knows how quickly texts become muddled when they are successively translated from Aramaic into Sanskrit and back again.

When it comes the turn of a termination codon to be read, two events can happen. Either the codon is recognized by a termination protein which attaches to the ribosome and detaches the peptide chain, or one of the cell's tRNAs associates with the codon. In the latter case, the tRNA sometimes remains attached long enough for its amino acid to be incorporated into the elongating protein so that the termination signal is crossed. When the termination codon is UGA, the tryptophan tRNA has two chances in 100 of succeeding. This is seen particularly well with the phage Qβ, which contains two genes separated by a UGA codon (note that termination codons are also called nonsense codons):

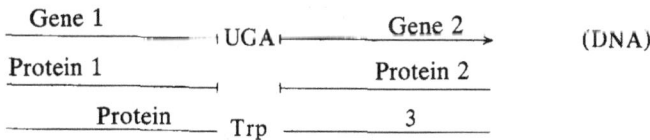

Under these conditions, two genes allow three proteins to be synthesized; protein 3 is made in small amounts and is indispensable for growth of the phage. A variant of tryptophan tRNA is known which reads the codon UGA very efficiently, crossing the termination codon with a frequency of thirty per cent; it has the same anticodon as the normal tryptophan tRNA from which it differs by a single base, which is far from the anticodon. It may be that the amino acid is positioned more favourably to be rapidly added on to the elongating protein before the codon–anticodon association is broken, thereby increasing the probability of forming a peptide bond. Such a tRNA which can read a termination codon is called a nonsense suppressor tRNA. Several are known which are more orthodox than the one which has just been described in that they carry mutations in their anticodons which allow them to form complementary associations with the codons which they suppress. There is also a class of mutant tRNAs (until now the mutations have always been found in the anticodons) which can read, with low but not negligible efficiency, a codon which does not correspond to amino acid which they carry. These are called missense suppressors and they illustrate well the indirect nature of the correspondence between codons and amino acids, since a change of the intermediary tRNA can change the meaning of a codon. Conversely, if a tRNA is charged with an amino acid which is not its own, the 'wrong' amino acid will be incorporated into the protein in place of the correct one.

A gene can be inactivated when one base too many is inserted in its sequence. During reading of the genetic message, when the zone containing the extra base is crossed, reading is shifted by one notch and the rest of the protein synthesized is unrecognizable. Correct reading can be restored if the insertion is compensated for by deletion of a neighbouring base so that most of the gene can be read correctly. Alternatively, correct reading can be re-established by combining three insertions; this produces one amino acid too many in the protein, but for certain enzymes this is not too serious. Crick, Brenner, Barnett and Watts–Tobin followed this line of argument and carried out one of the most beautiful experiments in molecular biology, which showed that the gentic code was based on codons of three letters (or multiples of three). There is another way of compensating for an insertion which succeeds in certain special cases: some mutant tRNAs can shift messenger RNA by one frame when they associate with certain codons.

Reading of the genetic message can also be affected by mutations in ribosomal proteins. Certain mutant ribosomes make many more mistakes than normal ones, while others make less. This situation is analogous to that of mutator and anti-mutator DNA polymerases, described in Chapter 6. Let us take a ribosome which makes mistakes. According to the old concepts, for example those found in Watson's *Molecular Biology of the Gene*, a ribosome makes mistakes because it is twisted or deformed at the codon–anticodon recognition site so that non-complementary associations are favoured while complementary ones are put at a disadvantage. However, ribosomes which make mistakes increase the reading efficiency of nonsense codons by nonsense suppressor tRNAs which have anticodons perfectly complementary to these codons. The 'structural' interpretation proposing distorted attachment sites thus leads to an impasse (Watson gets out of it by ignoring the fact that nonsense suppressors are more effective with ribosomes which make mistakes). This contradiction disappears when we abandon the structural point of view and turn instead to the energetic, probabilistic description of the genetic code. In many cases, a ribosome which makes mistakes is simply one on which peptide linkages are made too easily; errors increase and nonsense suppressors are more efficient.

Synonymous codons are used about indifferently to code for their amino acids in the genes of minor proteins. But highly expressed genes are constrained in their codon composition. Thus, proline is coded 38 times by CCG, and 10, 8 and 0 times by CCA, CCU and CCC respectively in the gene for the β subunit of RNA polymerase in *E. coli*. Proteins with a given function share a special amino acid composition, recognizable across the phylogenetic tree while – as shown by Grantham's team – the choices of synonymous codons in highly expressed genes say something about their cellular origin and little on what they code for.

Amino acid changes in proteins are mediated by mutations from codon to codon. Strong biases in the codon distributions will limit practically the feasibility of some amino acid substitutions. Reciprocally, if one kind of amino acid change is often desirable, the codon distributions of the two amino acids may end up mirroring each other. One could then dream of a theory that would consider all possible mutational pathways and show how protein and gene evolutions constrain each other. But Grantham's team's result – that protein and genes do cluster according to different rules – makes such a synthesis unlikely. Presumably, codon usage is mainly adjusted to the needs of the translation machinery.

As postulated by Garel then shown by him in the extreme case of silk biosynthesis, the cell synthesizes preferentially the tRNAs that read the most frequent codons. To optimize translation fully, the cell would make a messenger RNA structured enough to resist nucleases but loose enough to let ribosomes creep in. It would use successions of codons that slow down translation at some places to let the polypeptide chain fold properly, and successions that allow full speed elsewhere. It would try to reduce errors by keeping error-prone codons scarce and remote from error-enhancing contexts. Above all, it would choose the codons so as to avoid having the unfinished protein fall off from the ribosome (especially when close to the end). Single out any of the factors just listed, or any other plausible one, and you get one of the present or future theories of why particular cells use particular codons.

10 **The stereochemical hypothesis**

The stereochemical hypothesis retains a following among those who reflect on the origins of the genetic code. It is postulated, for example, that DNA used to govern protein synthesis without the participation of ribosomes, tRNAs or activating enzymes. DNA would have acted as an amino acid trap on which amino acids were bound side by side so that they could link spontaneously. If each zone of the DNA had a preference for a particular amino acid, the polypeptide sequence formed on the DNA would be a sort of translation of the nucleotide sequence. This idea, proposed by Caldwell and Hinshelwood (in 1950!) then by Gamow, was applied in the authors' minds to normal protein synthesis in contemporary cells. There is no waste in molecular biology and all shaky concepts are recycled towards the origins of life: the stereochemical hypothesis has become one of the major ideas in molecular evolution.

In DNA, the distance from one base pair to the next is close to 0.34 nm. Two successive residues in a polypeptide chain are separated by about 0.36 nm, almost the same distance, so a correspondence between DNA and polypeptides of one amino acid per base pair is plausible. This general concept has given rise to many variants. The declared objective of some is to resolve the mystery of the origin of the genetic code. Others are aimed at formulating general laws of recognition between proteins and nucleic acids which would come into play, for example, in the attachment of the repressor to its operator site. I have two major objections to make to the stereochemists. The first is that their solution is to a badly posed problem whose terms already include an underlying philosophy of specificity in biology which is not correct. Suppose that a beginner approaches two chess players. Ignorant of the rules of the game, he observes and counts the number of pieces which each player captures. A good statistician, he establishes a remarkable and highly significant correlation: the winner is nearly always the one who has captured more pieces. The object of the game, of course, is to capture one's opponent's king and no other piece, and the strategies used can only be understood in all their finesse in relation to this aim. In a similar way, despite the indisputable experimental evidence in support of it, the stereochemical hypothesis has missed the rules of the game, as I shall explain later. The second objection to the stereochemical hypothesis, in its most extreme form, is that it makes too much of coincidences. Before detailing these arguments, let us examine the proposals of the stereochemists.

For Gamow, the amino acid is attached in the wide groove of the double helix (*see* Fig. 2), making contact with four bases: a central pair, one base above it on one of the DNA chains and one base underneath it on the other chain. Put into retirement for 12 years or so, the stereochemical hypothesis reappeared when the genetic code had been worked out. Pelc and Welton proposed a model incorporating two recent advances. Knowing that the code goes by triplets, they proposed an interaction of one amino acid for every three bases, with an RNA chain replacing the DNA double helix. Next, the authors took into account the precise structure of helical RNAs which had just been worked out by the brilliant crystallographers of King's College, London, and which was later redetermined better and better under the successive names of RNA 10, RNA 11 and RNA 12. Using the most precise structural models, Pelc and Welton found that a structural complementarity existed between amino acids and their codons. The codons were constructed according to crystallographic results relating to RNA, differing from them on only one point: they were assembled back to front. Crick, who discovered the mistake, used it to discredit the stereochemical hypothesis again. But the idea was revived once more. After the iconoclasts Pelc and Welton had taken the first blows, their successors had more elbow room.

Shortly afterwards, young Dunnill wrote a short paper which, as time passes, seems to me to be the best of the series but which had no successor. Was Dunnill dissuaded from pursuing his research or did he really have only this message to leave us? The paper proposed a direct interaction between the amino acid and the anticodon loop of the tRNA. While in Pelc and Welton's version proline associated with its codon CCC, in Dunnill's version proline was recognized by the anticodon GGG. The anticodon loop in tRNA, seven nucleotides long, can form a natural pocket in which the amino acid could lodge. Since the conformation of the loop depends not only on the three central anticodon bases, but also on the four other bases, the interaction between the amino acid (proline here) and its anticodon (GGG) could only be anecdotal; it is the overall interaction of the loop with the amino acid which counts. The idea is plausible, physically, and most authors rallied to it, notably Crick and Woese. I shall discuss it again in the following chapter. Another youngster, B. R. Thomas, proposed that the amino acid was not recognized by the codon or anticodon alone but by the codon–anticodon sandwich; this proposal seems to have cut short his career in scientific research. Lacey and Pruitt changed one of the terms in the interaction and replaced the individual amino acid by the preassembled polypeptide. They suggested the existence of a privileged interaction between a regular polypeptide (poly-glycine) and a regular polynucleotide (poly-G). One of the codons for glycine is, as if by chance, GGG. Their model was presented as a central chain of poly-glycine in α-helical form, around which was wound the nucleic-acid chain in a configuration never seen previously, with three G residues per glycine. With the following attempts, made by biochemists or crystallographers of repute, the stereochemical hypothesis enters the era of respectability.

Adler and his collaborators proposed a code of recognition between DNA base pairs and amino acid in an α-helical polypeptide. Their aim was to predict interactions between DNA and the repressor of the lactose operon. In their model, the polypeptide is lodged in the wide groove of the DNA double helix. The contact points between the repressor and the operator are now known and are not those foreseen by Adler. But there are also β-pleated sheets in proteins, and if the wide groove of DNA is fit for the α helix, where should β sheets be lodged? Carter and Kraut thus proposed a direct interaction between β-pleated sheets and the wide groove of DNA as a basis for the origin of life. By good fortune, there were further possibilities to explore. As surmised by Rich, DNA is quite flexible and in some cases it turns left instead of right. The fashionable conformation in proteins is the Greek Key and it was then Mitsuyama's turn to propose a stereochemical model in which left-handed DNA is the proper lock to the Greek key.

What does the experiment show? The reality of interactions between nucleic acids and proteins is undisputable. The cell's nucleic acids are continually taken in charge by proteins which bind to them and stabilize or destabilize them, cut them or join them and modify them in other ways. Negative charges, carried by the phosphates in nucleic acids, attract directly the positively charged basic amino acids in proteins or indirectly the negatively charged acidic ones, through metal bridges. Aromatic amino acid residues intercalate between adjacent bases or base pairs. Owing to their relatively small size, they induce bends or kinks, but not much uncoiling. Finally, as Hélène and Lancelot sum up, 'all the polar amino acid side chains are able to form at least some hydrogen bonds with each of the four bases'. An amino acid can associate on its own with nucleic base. Saxinger and Ponnamperuma showed this by preparing gels with amino acids bound chemically to them and then filtering nucleotides on the gels: AMP, UMP, GMP, CMP and IMP. By measuring the quantities of amino acids retained, they deduced the association constants between amino acids and nucleotides. The hope of this work, published under the eyecatching title *Experimental Investigation on the Origin of the Genetic Code* was to gain evidence of strong preferences of certain amino acids for certain bases, but the results were disappointing. The nine amino acids studied (glycine, lysine, proline, methionine, arginine, histidine, phenylalanine, tyrosine and tryptophan) bind the five bases in roughly equal proportions, the purines a little better than the pyrimidines, with only two exceptions: glycine and proline show a slight preference for CMP over AMP. Peptides, which are half-way between amino acids and proteins, can associate effectively with nucleic acids and show evidence of a certain specificity, combining several effects. A good hundred or so peptides must have been synthesized in the laboratory and their interactions with nucleic acids of various sequences studied. Some specificities are apparent but there is scarcely any evidence of a code of nucleic acid–peptide interactions.

The philosophy which underlies these studies is that recognition between biological molecules is a matter of lock and key, of structural complementarity between molecules which associate. A little parable will reveal the weakness of this concept. Returning late to my hotel, I try to open the door of my room but the lock refuses to yield. I then notice that I had mistaken the door. Finding the right door, I open it without difficulty and admire the fine indentations on the key, imagining that the lock bears inside it the same complicated zigzags in reverse. The next day, the chambermaid opens all the doors on the corridor with calm assurance and an almost magical key: the pass-key. It might be thought that the pass-key is even more elaborately indented than mine since it opens all the doors. But no, it is a smooth and everyday key. The reason is simple. The important thing for the hotel management is that the guest cannot enter a room other than his own. The design of each key obeys two conditions: to be able to open the corresponding lock but not the others. The first condition is easily met: the pass-key, a jemmy or a screwdriver do the job very well. It is the second condition which is difficult to satisfy and the locksmith has to concern himself with games of locks and keys which present the fewest possible cross-interactions. The same goes for the cell. Recognition is the result of a choice between a finite number of possibilities, obtained by superimposing on a general principle of interaction a group of essentially negative measures to eliminate undesirable possibilities. We have seen this notion emerging in relation to codon–anticodon recognition. The same generalization is valid for recognition between tRNAs and activating enzymes. In the cell, the activating enzyme for phenylalanine is capable of charging the tRNA specific for phenylalanine and that one alone. But if it is presented with totally 'unknown' tRNAs from a distantly related species, the same enzyme will be capable of acylating a good third of these tRNAs (we will discuss this again in Chapter 15). The specificity of the enzyme thus resides not in its capacity to associate with its own tRNA, but in the fact that it rejects the tRNAs of its own cell which correspond to another amino acid. When confronted with unknown tRNAs, not knowing to reject them, it acylates them.

When a mathematician is given four equations to solve, which contain only three unknowns, he does not expect to be able to find the overall solution since the unknowns have to obey too many conditions at once. It seems to me that, in the same way, the stereochemical hypothesis asks too much of amino acids and nucleic acids. The four bases already have remarkable properties in their capacity to form stable, symmetrical and geometrically equivalent pairs. Prebiotic synthesis of purines seems easy but that of the pyrimidines, C especially, is much more difficult. The simultaneous existence of two purines and two pyrimidines doubtless required a very special environment.

To expect in addition to this, that each sequence of three bases distinguishes a single amino acid from a list of twenty is surely too much for the same molecules. Some supporters of the stereochemical hypothesis entitled their book on the origins of life *Biochemical Predestination*. The title suggests connotations of that stream of biological thought which is characterized by belief in the pre-established harmony of molecules and perhaps in the Supreme Craftsman who designed them so well.

11 The origin of the genetic code

The genetic code, centre-piece of cellular logic, is linked in one way or another with everything which occurs in the cell. When we examine cellular function we can find many tracks by which one can hope to reach the origin of the code. Having chosen his starting point, each author also has his own way of classifying theories on the origin of the code. I distinguish two approaches, that which considers the code as a dictionary, and, much more interesting, that in which the code is understood as a process. The principal concern of the theoreticians in the 1960s seems to have been to explain the structure of the genetic dictionary. Why does UUU correspond to phenylalanine while UGG codes for tryptophan? Monod poses the alternatives thus: '(a) The structure of the code is explained by chemical reasons, or more exactly stereochemical ones. . . . (b) The structure of the code is chemically arbitrary; the code as we know it results from a series of random choices which have enriched it little by little.' Nowhere is the question posed as to the origin of the coding *process*.

The first answers to the question of the genetic code's structure were – as might have been expected – the code is as it is because it is the best! Thus Conrad imagines that several genetic codes operated when life was beginning; the first code to succeed in making a useful protein, for example a DNA polymerase, gained an advantage over the others which made uninteresting proteins from the same messenger RNAs. The idea was taken up again by Eigen. According to Sonneborn, most mutations in a gene have a harmful or even lethal effect, so the dictionary would have evolved so as to minimize the effects of mutations; starting from whatever dictionary, evolution would have led to a situation in which the codons for the same amino acid would be regrouped into compartments and the compartments containing chemically similar amino acids would be close together. In these conditions, point mutations would have a high probability of transforming one codon of an amino acid into another codon of the same or a related amino acid. As Woese showed, Sonneborn's idea is logically incoherent. At the moment when the meaning of a codon changed, all the genes containing the codon would produce altered proteins. The premise of Sonneborn's argument – false in my opinion – is that most mutations are lethal; it is then obvious *a fortiori* that a change in the meaning of a codon leading to tens or hundreds of simultaneous mutations would be hyperlethal. One falls into the very trap which one wanted to avoid.

Crick considers that, for structural reasons, the degeneracy of the code has existed since the beginning in its present form. Because of the laws of loop formation in nucleic acids, primitive anticodons were in a special configuration allowing perfect recognition of the codon's first two positions and less good recognition of the third. If the wobble rules applied in every detail to primitive codon–anticodon recognition, the number of codable amino acids could not have exceeded 32 (according to wobble, every anticodon capable of pairing with XYC must also be able to pair with XYU, etc.). Crick proposes that, in the beginning, only four or five amino acids were coded for. New amino acids would have infiltrated into the code later by usurping a rarely used codon from a chemically similar amino acid. The idea of evolution of the code by infiltration of amino acids was taken up again, from another viewpoint, by Dillon then by Tze-Fei Wong. We may assume that at first only amino acids which were easily synthesized in the prebiotic soup were coded for. Later on, the evolution of the code would reflect that of metabolism: if a new pathway allowed a new amino acid, for example tyrosine, to be made from an old one, phenylalanine, then tyrosine would enter the code, usurping some phenylalanine codons.

Woese maintains that every primitive translation was necessarily imprecise, notably because of the difficulty of distinguishing between amino acids of closely related structures such as valine and isoleucine. Initially, each codon, GUG for example, would have designated several amino acids at the same time – say valine, leucine and isoleucine. There would be a certain probability of incorporating any one of these amino acids in response to the codon GUG. Several different proteins would be synthesized from the same gene, or as one says, the translation product would be a 'statistical protein'. Later on, the meaning of the codons would have been made more precise, like an image viewed with a camera which is blurred at first then becomes clearer as the focus is improved. Going a little further, Orgel proposes that, in the beginning, the code distinguished between only two classes of amino acids – the hydrophobic ones and the hydrophilic ones. The idea of evolution from blurred to clear may be transposed to the level of codon–anticodon recognition. If it is imprecise, a codon A is read by several potential anticodons A_1, A_2, etc. But each potential anticodon can read several codons which will in some way be 'cousins' of A. If these comings and goings continue until no more new codons come into play, a first list of codons will result. In this way the sixty-four codons can be separated into a small number of lists. At first, there would have been only one amino acid coded for per list. Subsequent improvement of anticodon loops by chemical modification, and in particular the use of inosine, would have permitted further subdivision of the sixty-four codons into lists and increased the number of amino acids coded for. This idea, developed by Claverie, was taken up again by Jukes, without reference to its origin, and Barricelli, the pioneer of cellular automatons.

The systematic character of the degeneracy of the code in position 3 suggested to Jukes that initially codons were only two bases long. Crick opposed this idea with the 'Principle of Continuity' which Orgel expressed thus:

'I shall be guided by a Principle of Continuity which requires that each stage in evolution develops "continuously" from the previous one. It is very difficult to see how a totally different biological organization could have undergone a continuous transition to the nucleic-acid–protein system with which we are familiar. Thus, at least until such time as a reasonably detailed model of a novel system is suggested and a means for its evolution into the present system is proposed, I feel justified in supposing that certain features of the contemporary genetic system emerged very early in the development of life.'

And here is how Crick goes from this general principle to refute the idea of a primitive code using codons with two bases: 'It might be argued that the primitive code was not a triplet code but that originally the bases were read one at a time (giving 4 codons) then two at a time (giving 16 codons), and only later evolved to the present triplet code. This seems highly unlikely since it violates the Principle of Continuity. A change in codon size necessarily makes nonsense of *all* previous messages and would almost certainly be lethal.'

This argument is not irrefutable and can be challenged by the idea of evolution by take-over developed by Cairns-Smith. For example, consider a cell which uses a doublet code. A very minor class of ribosomes arises by mutation which uses the cell's tRNAs and translates a small number of messenger RNAs in groups of three bases. We thus have a parasitic translation in triplets in an organism whose major mode is translation by doublets. It may then be imagined that the parasitic mode gradually increases in importance. Working on their own messenger RNAs, the new ribosomes handle a larger and larger proportion of protein production. The old ribosomes are used less and less and eventually disappear. Here we see how two different modes of organization can succeed one another. The first does not evolve into the second but creates favourable conditions for the appearance of the other which later supplants it. Cairns-Smith considers that the present organization with nucleic acids, proteins and the code may have been preceded by a radically different form of life based on clay-like moulds. Without going as far as this, it is possible that nucleic acids may have been preceded by chemically simpler polymers which were also able to form complementary structures.

This brief sketch leaves us unsatisfied. We have discussed the direction in which the code could have been modified once the translation apparatus was in existence. But this tells us nothing about the appearance of this apparatus in the first place. I will now deal with this subject by presenting three trends which lead to different experimental verifications, then close on a joyful note with the theory of Eigen.

It is probably Woese, more than any, who launched important ideas concerning the origin of the genetic code in his articles and his **prolific** writings have been, without doubt, a mine of inspiration for much theoretical work. He was able to put his finger, at just the right time, on several important notions: the stereochemical hypothesis, systematic degeneracy of the code, chemical modification in the anticodon loop and regularities in the dictionary. He very clearly affirmed the primacy of the process over the dictionary and was celebrated in the American press as the discoverer of a new form of life (see p. 27). His distinction would have been greater had he pursued his best ideas or premonitions on the genetic code and the origins of life more systematically to the point at which they would have found solid contact with experimental reality.

In his now classic book on the genetic code, Woese presented a picture of the chemical context in which the code might have begun: polypeptides, nucleic acids, and the small molecules of the prebiotic soup. What were the first nucleic acids like? They were certainly quite irregular since evolved enzymes are necessary for precise replication of nucleic acids or at least to ensure regular chain formation – base–sugar–phosphate–base, etc. Nevertheless, certain regularities could have appeared in the absence of enzymes. Woese notes, among physical principles of order, that guanine monomers tend to form helical stacks in solution. If this occurs spontaneously, it may facilitate formation of homogeneous chemical bonds between the monomers and so lead to chemically regular nucleic polymers. Primitive polypeptides, with random sequences, could have possessed weak catalytic activities which were not highly specific. If polypeptide synthesis were performed on certain crystal surfaces, it might have allowed a certain structural homogeneity to be established, for example preferential use of L- or D-amino acids according to the asymmetry of the crystal. Metal ions, around which polypeptides could coil, may thereby have been a factor in sequence selection. The combination of a metal ion and a polypeptide might have allowed certain photochemical reactions to occur.

How could a translation apparatus develop under these conditions? Let us reason in reverse; what can we take away from the present apparatus? According to Woese, the most plausible simplification consists of doing without activating enzymes. These enzymes are useful only if they are very specific and therefore synthesized with precision. It is difficult to see how a very primitive translation system could have synthesized in a sufficiently precise way enzymes which have to recognize, without errors, both tRNAs and amino acids. Woese, and practically everyone after him, considered that the primitive tRNA itself had the capacity to recognize the amino acid. Like Dunnill, Woese imagines that the primitive tRNA contained a loop (later to become the anticodon loop) which could bind one or several amino acids more or less specifically. Once fixed at the anticodon site of the tRNA, the amino acid would be transferred to a second site where peptide synthesis would be carried out. tRNA, notes Woese, is a small molecule but is particularly complicated in comparison with the other RNAs in the cell; sometimes one base in six is something other than A, U, G or C. This suggests that tRNA may have had quite a complex role: that of a primitive enzyme. A second simplification: let us suppress messenger RNA! Woese conceives of primitive tRNA as an enzyme producing polypeptides in a non-coded way. When the first amino acid has been recognized by the anticodon loop then transferred to the catalytic site of the tRNA, the second amino acid can attach to the anticodon loop, be in its turn transferred to the catalytic site and form a peptide bond with the first amino acid, and so on. Primitive RNA would thus have produced homogeneous proteins or ones in which the average amino acid composition was the same along the chain. A second stage in the evolution of the synthetic process would be to make peptide bonds using two tRNAs placed side by side. Primitive ribosomes would have had no other use than to provide a surface for attachment of tRNAs; relatively unspecific complexes between RNA molecules and polypeptides could have done this. Later, peptide synthesis on the surfaces of primitive ribosomes would have become more and more specific through the intervention of messenger RNAs which preferentially stabilized the binding of certain tRNAs to the ribosome. Thus, the ancestor of the translation machinery would have carried out non-coded polypeptide synthesis; correspondence between nucleic and protein sequences would have been a later development.

If the primitive protein synthesis machinery made uninteresting products, it is difficult to see how they could have been the source of later improvements. An indirect way of testing Woese's ideas, therefore, is to examine statistical peptides, which are liable to have been made by primitive tRNAs, and see if they can have interesting catalytic properties or not. Fox has shown that amino acids can be polymerized to form 'proteinoids' by heating them above 100°C. Proteinoids tend to aggregate and form extended molecular assemblies: 'microspheres'. Attempts have been made to find catalytic activities in proteinoids. The systematic work undertaken by Dose revealed some activities, including hydrolysis of ATP, but there were no spectacular results.

Another major figure in molecular evolution, Orgel displays a temperament quite unlike that of Woese. As a young chemist, he became famous through his work on transition metals and could have spent the rest of his days embellishing his theory, which would no doubt have been to his advantage for academic honours and distributions of medals. Through his contact with Lord Todd, in whose laboratory he rubbed shoulders with two pioneers of nucleic-acid chemistry: Michelson and Khorana, then through contact with Crick, he gained a deep knowledge of molecular biology and from then on devoted most of his efforts to the problem of the origins of life. With determination and method Orgel, in his laboratory at the Salk Institute, perched above the most beautiful nudist beach in California, is attempting to reconstitute the part of prebiotic chemistry which most interests the molecular biologist, the syntheses of proteins and nucleic acids. The hope is to approach gradually the level of complexity at which the two polymers would interact in ways prefiguring the coding relations. He was one of the first to recognize the difficulty of achieving and maintaining cellular accuracy and developed the concept of error catastrophe which will be discussed in the final chapter. Like a well-educated Briton, he is loath to go down into the arena to defend his less conventional ideas and prefers to keep them for discussion with his peers. He thus appears more orthodox than he is and so risks losing the support of the younger generation.

According to Orgel, the primitive genetic code would have established correspondence between messenger RNAs with alternating purine–pyrimidine–purine–pyrimidine, etc. and polypeptides with alternating hydrophilic and hydrophobic amino acids. Firstly, he says, prebiotic replication of nucleic acids containing several purines in succession must have been difficult – we saw this in Chapter 8. Alternating nucleic acids, which never contain two consecutive purines, must therefore have replicated much better than non-alternating ones and have accumulated in the prebiotic soup. What would happen, with the present code, if an RNA with alternating purines and pyrimidines was translated? We would obtain a protein in which two types of amino acid alternated, those with codons of the type purine–pyrimidine–purine, and those with codons of the type pyrimidine–purine–pyrimidine. Now, Orgel points out, the first are mostly hydrophobic and the second hydrophilic. A chain formed from alternating hydrophilic and hydrophobic amino acids should, Orgel postulates, have a tendency to fold up into a β-pleated sheet, which is the preferred conformation of proteins, especially those which appear to be ancient like ferredoxin. Because of the alternation, such a β-pleated sheet would expose hydrophilic amino acids at one side and hydrophobic ones at the other and would thus be an ideal material for forming a membrane separating an exterior aqueous medium from a much less hydrated interior medium. The structure of the dictionary thus seems ideal for ensuring translation of an alternating messenger RNA into membrane protein or more generally into β-pleated sheet polypeptide.

The third point in support of Orgel's theory comes from codon–anticodon recognition. First, reading of codons is blurred in position 3, a little clearer in position 1 and more precise in position 2. Secondly, there is a good correlation between the nature of the second base in the codon and the properties of the amino acid coded for. Let us suppose that primitively there were two classes of tRNA, one set bearing hydrophilic amino acids and having a pyrimidine in the middle position of the anticodon, the others bearing hydrophobic amino acids and having a purine in the middle of the anticodon. If we assume that reading of codons was blurred in positions 1 and 3 but precise in position 2, then translation of alternating RNAs must have given rise to alternating polypeptides. The idea that alternating polypeptides have β-pleated sheet structures is a chemist's hypothesis. All alternating polypeptides synthesized in the laboratory have β-pleated sheet structures and there seems to be a slight tendency towards alternation of hydrophobic and hydrophilic amino acids in the β-pleated sheets of known proteins.

Later, Orgel and Brack (who studies alternating polypeptides in Orleans) made an interesting addition to this theoretical edifice. Comparative analysis of the active sites of proteins reveals a quite frequent conformation, the β-barrel (Fig. 11, p. 32). Zigzags form the sides of the barrel and the loops meet on one side or the other. The junction of several loops forms an ideal site for binding a substrate and performing catalytic activity. Overall what is necessary to obtain a protein endowed with catalytic activity? First, a general β-pleated sheet structure. Next, folds with well-adjusted lengths so that the loops join up. Lastly, amino acids capable of catalytic activity, such as serine or histidine, must be present in certain strategic positions in the loops. With this view, it is relatively easy to understand how rudimentary coding might be sufficient to specify catalytic proteins.

These primitive proteins could be simulated experimentally. First, one would synthesize chemically an alternating polypeptide (which one does not expect *a priori* to show catalytic activity) and then modify it chemically by attaching active groups: imidazole, acetyl or others. Nothing has been achieved in this way because nothing has been attempted. But we can be optimistic since, for several decades now, organic chemists have been making active catalysts according to a similar general principle: active groups are hooked to a polymer which provides anchoring points; all the skill lies in choosing the polymer which will allow the groups to be positioned at ideal distances and orientations for catalysis. For example, Klotz prepared a synthetic resin, polyethyleneimine containing dodecyl groups ($C_{12}H_{25}-$) to which he attached imidazole groups. The resulting polymers catalysed the hydrolysis of 4-nitro-catechol sulphate 100 times faster than the enzyme specialized for this kind of task, aryl-sulphatase.

I arrived at my ideas on the genetic code by reflecting on an imaginary experiment in 'evolutionary regression'. Let us consider a modern bacterium in which all cellular processes are accurate. Then let us lower the level of accuracy. Beyond a certain threshold the bacterium will perish through accumulation of errors. Now, the great precision of the present has been acquired through evolution. We must assume that more-primitive bacteria were able to grow and multiply with quite low levels of precision which would be lethal for a present-day bacterium. What did they have which was different? The idea which seemed plausible to me was that their proteins were shorter. Let us suppose that, in protein synthesis, one amino acid in fifty is incorporated by mistake, on average. Then proteins fifty amino acids long will have, on average, one incorrect amino acid in their sequence. More exactly, 36.4 per cent of proteins would be an exact translation of their gene, 37.2 per cent simple variants and 18.6 per cent double variants containing two incorrect amino acids, etc. For a protein 200 amino acids long the distribution would be: only 1.76 per cent standard copies, 7.2 per cent simple variants,

14.6 per cent double variants, etc. Although it cannot be proved, I consider that the second situation is clearly less favourable than the first. To be sure of this, we would need two enzymes which carried out the same function but were of different lengths and systematically mutate them. If proteins become shorter and shorter the further back in time we go, then the primitive ancestor of the translation machinery must have made short peptides and not proteins. But then, a genetic code would not have been necessary. A small collection of short peptides with well-defined sequences can be produced by rudimentary synthetic methods without ribosomes or tRNAs. Let us look at the problem from this end and go back in the opposite direction. I am supposing that we have primitive synthesis, catalysed by a crystal or whatever, of a defined dipeptide, for example methionine–tyrosine. Later this synthesis becomes complicated by one step; a tripeptide is formed: methionine–tyrosine–valine, the bond between the second and third amino acids being facilitated by the presence of an oligonucleotide, let us say AGCG. There is no coding relationship, just a coupling between two events: binding of a cofactor AGCG and addition of a valine to the dipeptide methionine–tyrosine. At a third stage a tetrapeptide methionine–tyrosine–valine–histidine, or the same with glutamine as the fourth amino acid, is synthesized. As for the tripeptide, synthesis would depend on the presence of an oligonucleotide, the nature of which would determine which of the two tetrapeptides were made. We still have no genetic code, but peptide synthesis with 'options' on the fourth position. We now see the distance which separates this synthesis from one using a code: the system with options has to become repetitive so that, from the fifth amino acid on for example, the same regular process is reiterated allowing the next amino acids to be put into place. Later, the regular iteration can be made to start right from the beginning of the chain. To sum up, what is fundamental in the genetic code, from my point of view, is not linear correspondence between a messenger RNA and a protein but the existence of an 'elongation cycle'; a reiterative process which causes the nth amino acid and the $(n + 1)$th to be added on in exactly the same manner.

Accessorily, this led me to reflect on the three-dimensional structure of tRNA and its evolution. We can imagine a primitive tRNA made up of *two* pieces: a short oligonucleotide to which an amino acid is attached non-specifically and a hairpin – that is, the end of a nucleic chain folded on itself and forming a loop in its middle:

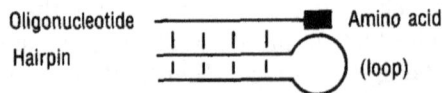

We know that quite often a nucleic acid double helix can be associated with a supplementary filament which comes to lie in the wide groove of the double helix, forming a three filament structure. The primitive tRNA would have been formed from the association of the hairpin with the oligomer carrying the amino acid, energy for the interaction being derived principally from formation of the triple helix between nucleic chains. The amino acid, attached to the oligomer, would come into contact with the loop, which would influence its position in space for various reasons (attractions due to opposite charges, repulsions, steric hindrance). This primitive tRNA, without an anticodon, combines a general principle of attraction without specificity (between nucleotide chains) with a principle of specific positioning not requiring attraction (interaction between the amino acid and the loop). From here it was possible to conceive of successive enlargements of the molecule, according to the order of events shown in Figure 16.

One experimental counterpart of the theory which I am proposing is the search for catalytic activities in short peptides. Organic chemists have long been interested in peptide catalysts and they have invented several capable of hydrolysing the choice substrate for this kind of study, paranitrophenylacetate. This compound hydrolyses spontaneously in water with appreciable speed so it was a sure bet that catalysts would be found without much problem for this easy reaction. In fact, imidazole can accelerate the reaction by itself; numerous peptides have been made of $n = 3$–6 amino acids long which accelerate hydrolysis of the substrate by a factor equal to $30 \times n$ that of imidazole alone. A quite simple peptide, glutathione, acting in conjunction with a metal, molybdenum, can catalyse a reaction analogous to the fixation of nitrogen: the reduction of hydrazine to ammonia. The glutathione–molybdenum complex reduces acetylene to ethylene with a speed comparable with that of enzymes. Lys-Trp-Lys, a tripeptide synthesized in Helene's group, proved to be capable of binding to damaged sites of DNA which had lost a purine nucleotide, and break there the DNA backbone, a task for which the cell normally produces a special endonuclease. In a more ambitious work, Gutte and co-workers designed a 3D-model for a polypeptide capable of interacting with nucleic acids. The model included two antiparallel β-zigzags forming a sheet, connected to an α-helix of similar length. They synthesized a 34-residue polypeptide whose conformation was predicted to agree with the model. The polypeptide did bind to single-stranded DNA and did hydrolyse it at preferential sites (the 3′ side of C residues) when present as dimers. The monomer showed little hydrolytic activity. But Gutte's pleasure at this result was not complete since his artificial enzyme refused to adopt the conformation it was designed for.

The future of this branch of science, it seems to me, lies in the use of selective methods. When the biochemist is led to postulate the existence of a certain enzymatic activity in the cell, he prepares an extract containing all the soluble proteins from the cell and then tries, by trial and error, to find conditions in which the postulated activity shows up. Once he has put his finger on it, he tries to separate the active principle from uninteresting proteins and gradually purifies an enzyme which initially was invisible. This approach may be extended to peptides: one can synthesize mixtures of

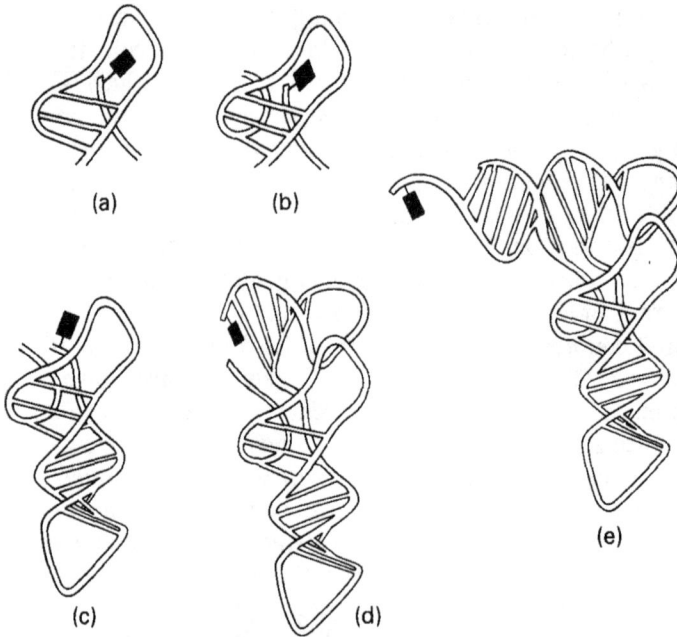

Figure 16   Three-dimensional evolution of transfer RNA. The spatial structure of transfer RNA, as determined by the teams of Rich, Kim and Klug using X-rays, is shown in (e) in a representation similar to that of DNA in Figure 2 (c). The following hypothesis about the evolution in three dimensions of this structure makes it seem less mysterious. I suppose that initially the ancestors of transfer RNAs were formed from two separate molecules: an oligonucleotide carrying an amino acid (black rectangle) and a segment of double helix with a loop. The energy for the interaction comes from binding of the oligonucleotide in the wide groove of the double helical segment. The loop allows the amino acid to be positioned (a). In (b) positioning of the amino acid is made more precise by a turning movement of one of the strands. In (c) the two molecules become one and the anticodon loop appears. The transition from (c) to (d) is the delicate step of the model. However, the new section can be added on without displacing the amino acid too far from its previous position. Finally, the tRNA tail elongates progressively to attain its present size. Adapted from a similar model proposed before the 3D-structure of tRNA was known (ref. 194).

peptides 'at random' of varying amino acid composition and look for catalytic activity in the presence or absence of metal ions. If an activity is found, the mixture can be purified and the interesting activity isolated, or new mixtures can be synthesized – taking into account the information gained – which may be enriched in the activity. We can go further and dream of constructing a machine which would couple peptide synthesis to the performance of a reaction which we are trying to catalyse so that it would selectively amplify the synthesis of peptides which catalysed that reaction. Is this possible? In 1972 Orgel organized a small colloquium on this theme.

Half-way between the sifting method inspired by biochemistry and gene-tics, which I propose, and the peptide-selecting machine which Orgel envisaged are various semi-selective methods. For example, a means of obtaining a peptide catalyst capable of acting on DNA would be to prepare a mixture of peptides and filter them through a resin containing DNA which would retain peptides having affinity for it. After this selection, the peptides could be subjected to a sifting procedure to detect either destabilizing cutting peptides, or stabilizing, cutting ones. I consider that selective methods for producing peptides at will with a given activity will eventually be of more benefit in medical applications than the production of authentic enzymes by transplantation of genes. There are two obstacles to achieving success in these projects. First, those who dispense research funds, as much in France as in the United States, do not look too kindly on anything to do with the origin of life. Second, the organic chemists who synthesize peptides believe that they know enough to invent the most active peptides: resorting to selective methods, which require less knowledge, seems degrading to them.

The origin of the genetic code is still a very open problem and speculation about it will continue for a long time. The aim of determining exactly how the code appeared on earth seems excessive to me. To produce a rudimentary coding system in prebiotic type experiments, even one very far removed from the present code in its working, would be a great success. Considering the areas of doubt, it is surprising to learn that Eigen appears to have demon-strated by calculation that the appearance of the genetic code was not only possible in the prebiotic soup, but even obligatory – this being based on the irreversible thermodynamics of Prigogine and Glansdorff. Recalling Eigen's calculations, Prigogine returned the compliment: 'If the theory of Eigen is confirmed, it will certainly be the concern of a field of fundamental research, since for the first time, a highly organized state, corresponding to a genetic code, would emerge from physical laws'. Further on in the same article, he is still more affirmative: Boltzman's Principle of Order, dissipative structures and code are the links of a chain which leads from thermodynamic equili-brium to biological order. And he concludes beautifully: 'of course, on other planets the forms taken by life will be able to differ since dissipative structures conserve the memory of the fluctuations which gave them birth'. But it seems to me not unreasonable to think that the life phenomenon is as foreseeable as the crystalline state or the liquid state. All this rests on Eigen's calculations on cycles and hypercycles of chemical reactions which Eigen presents thus: As a consequence of such instability, the nucleation of this functional correla-tion* (we may call it the origin of life) turns out to be an inevitable event – provided favourable conditions of free energy flow are maintained over a sufficiently long period of time. The primary event is not unique. Universality of the code will result in any case as a consequence of nonlinear competition.

* 'Nucleation of functional correlation': the formation of a code connecting nucleic acids and proteins.

The genetic code is no longer universal. Non-linear competitive hypercycles have survived, which proves their adaptive value.

It is my duty to say that I do not agree. Eigen and Prigogine both have the Nobel prize which they deserved, one for the development of methods of ultra-rapid kinetic analysis of chemical reactions, the other for his work as a thermodynamicist. Under the cowardly morals of today, they are unlikely to be contradicted, even when they deal with matters that are outside the domain in which they gained their reputation. Unfortunately, the weight of two or three influential academics is sometimes sufficient to seal the fate of a subject over a whole country. If the French took 50 years to accept Newton and almost twice that to accept Darwin, and if the origin of life is a taboo subject in the C.N.R.S.,† is this because the French on the whole are stupid or is it because French university institutions were or are in certain cases led by a small number of scientists with sufficient weight to block the development of a whole discipline?

Let us get to the facts. In a sixty-page paper in the journal *Die Naturwissenschaften*, Eigen presents a mathematical theory of the origin of life. In it, DNA molecules reproduce, mutate and compete; this type of treatment has been used for a long time in population genetics. Dealing with the code, he imagines a prebiotic soup containing enormous quantities of activating enzymes, tRNAs and messenger RNAs. Let I be an amino acid and J the anticodon of a tRNA. Suppose that for each pair IJ there are activating enzymes which can attach the amino acid I to the tRNA with the anticodon J. If we make a lottery with messenger RNAs, tRNAs and activating enzymes, will we have a coherent whole? The whole will be coherent only if, when the messenger RNAs are translated by activating enzymes and tRNAs produced by the lottery, we obtain activating enzymes of the same specificity as the initial enzymes. Then the system is self-maintaining; it lives. Certain particularly fortunate draws will produce a coherent system. Eigen calculated that the probability of such draws occurring was reasonably large.

Let us get back to the prebiotic soup, populated with all the elementary components of translation systems, and suppose that in addition the soup contains an immense number of cells: objects delimited by membranes, endowed with autonomy. In Eigen's treatment the word cell is delicately replaced by the expression 'element of volume'. The probability calculated by Eigen is that of finding a coherent translation system all in the same 'element of volume'. The element of volume in this calculation is a watertight compartment, otherwise the coherence of the system would be transitory with tRNAs quickly diffusing and mixing with neighbouring ones. Eigen has clarified this point a few pages earlier: 'The system, after nucleation, has soon to escape into a compartment.' From here, Eigen arrives at the conclusion

---

† Translator's note: the C.N.R.S. (Centre National de la Recherche Scientifique) is the French Government body responsible for organizing and funding much of French science.

that the necessity for membranes can be deduced from his theory. Admire: 'The "individualization" of the hypercycle – which thereby becomes a truly "self-reproductive" system – has to be seen in connection with "compartmentalization". Neither "individualization" nor "compartmentalization" are inherent properties of the hypercycle – as are, for instance, the other properties mentioned above. However, where they occur after nucleation, they may offer a selective advantage and therefore are inevitable evolutionary consequences of the hypercycle.' Inevitable consequences: Eigen is not afraid of words.

Let us sum up: if in the prebiotic soup we have membranes, tRNAs, messenger RNAs and activating enzymes, then the probability of the appearance of the genetic code is not zero! Isn't this just what would have been expected?

12 Sequence space

This is just a little metaphor, but it provides a means for rethinking the whole of molecular evolution and making visible the intimate connections between the great questions. First, let us see how these things used to be considered.

In the Darwinian view, every protein sequence is the result of a sifting by natural selection through all possibilities. Let us try to calculate the number of attempts which were necessary to produce a protein as simple as cytochrome c, only 105 residues long. There is one chance in twenty of synthesizing by a random process a protein which begins with the same amino acid as cytochrome c, then one chance in twenty that the second amino acid will be the correct one, and so on. There is one chance in twenty to the power 105 of making this sequence by trying all possible variants of any given starting sequence. To have a reasonable chance of making the cytochrome c sequence, it is therefore necessary to make in the order of $(20)^{105}$ sequences. This is a fabulous number: if one million different sequences were synthesized every second in each cubic millimetre of ocean for 3000 million years on a million different planets, the chances of finding the right sequence would still be ridiculously small. So, according to this argument, Darwinian evolution would not have had the time to produce even one cellular protein. Naturally, several authors have tried to refute the argument.

One unhappy attempt is that of Eigen and Schuster who presented the following model. Suppose that we want to form a particular sequence, for example: ABCABDABE from a given squence such as CLBBDBDAE by allowing this sequence to reproduce and mutate by chance. Let us now introduce a selection factor. Every time two identical letters are found in the same position in the final sequence and the evolving one, we attribute a selective advantage of ten. In the model, the sequence CLBBDBDAE has an E in position 9 in common with ABCABDABE and so will replicate 10 times faster than every mutant which does not have a letter in common with it. The mutant ALBBDBDAE, with two letters in common with the final sequence in homologous positions, would replicate 10 times faster again than CLBBDBDAE and therefore 100 times faster than a sequence with no letter in common with the final sequence. By simulating this process of mutation and selection in the computer (the great IBM wizard always impresses biologists), Eigen and Schuster showed that the initial sequence evolves very rapidly into the final one. Mutation guided by selection goes straight to the target. Unfortunately natural selection is the expression of competition between species based on their present merits and not on what they might become in the distant future.

A more physico-chemical argument is given by the information scientist Yockey. Amino acids can be divided into equivalent classes according to their properties: aspartic acid and glutamic acid form such a class; they are both acidic and the replacement of one by the other in a protein should not change its activity greatly. Similarly, one can substitute an aromatic amino acid for another without affecting too much the properties of the protein. Yockey sees a protein as a sequence of symbols, each symbol representing a particular class of amino acid. The important thing would then be not to make a particular cytochrome *c* sequence, but to obtain one of the very many sequences which conform to the model symbol sequence. Despite this adjustment, the probability of obtaining such a sequence by chance remains ridiculously low, and Yockey serenely concludes that Darwinian evolution is impossible.

The metaphor of Sequence Space, proposed by Maynard Smith in 1961, allows us to reason in a much sounder way. Let us represent every possible protein sequence by a point in space, which we will call protein space. For simplicity (mathematicians will make the necessary correction themselves) each protein is represented by a point in a plane; two proteins with closely similar sequences are close neighbours in the plane. Evolution of sequences can be seen as a 'march' in protein space from one point to another. What is the terrain like? We introduce relief into this space so that it now has three dimensions. If the protein is totally inactive, we have a peak; if it is active as a catalyst, we have a hole. The depth of the hole will vary according to the activity of the protein. The holes and peaks will form a landscape. For example, in a certain zone of protein space there might only be peaks, which would form a great mountain range. In other places, there could be many contiguous holes, which would form basins or river-beds. Many of the questions posed by molecular evolution can be expressed in these terms: what is the protein landscape like? For the ultra-selectionist, the protein landscape would bristle with peaks surrounding some rare holes, which would have difficulty meeting up. For the neutralists, on the other hand, the protein landscape undulates very slightly containing vast basins and round hills, but never varies far from the mean level.

Let us now consider two activities, for example a DNA polymerase and an RNA polymerase, each occupying its hole. Is there a channel for communication between the two holes? Orthodox molecular evolutionists used to answer no, and strongly emphasized the necessity of mutating a gene over a long time period before it could take on a new skin.

Maynard Smith, however, envisaged a protein landscape allowing gentle transition from one activity to another. As an analogy he uses those word games in which words are changed into others, one letter at a time, and each intermediate word must have a meaning, for example: WORD–WORE–GORE–GONE–GENE. The most recent results on bacterial evolution (the following chapter) show Zuckerkandl and Ohno to be wrong and rather favour the means of Maynard Smith. Let there be two proteins A and B which have distinct activities in the cell but have biochemical similarities. The specificity of A or B is never absolute. Enzyme A can carry out the reaction normally catalysed by B but with very low efficiency, let us say one ten thousandth that of B. A point mutation turns enzyme A into A′; slightly less active with its natural substrate but ten times more active in the traditional reaction of B. Thus type A and type B activities potentially coexist in each of the two enzymes A and B. A series of point mutations gradually reinforces the minor activity.

The time is now ripe to raise two questions. What is the appearance of the 'contour lines' about a given activity? And what is the probability that we will find an interesting activity at any point chosen at random in protein space?

The most direct method for showing what the contour lines around a given activity are like consists of artificially varying sequences, in a systematic manner, and studying the properties of each variant. In this way, in 1968, Langridge derived fifty-six proteins from β-galactosidase. The method consisted of first introducing a nonsense codon at a predetermined position in the β-galactosidase gene then a nonsense suppressor tRNA, a carrier of the amino acid serine. This produced a variant of β-galactosidase in which serine had replaced the amino acid specified by the codon which had become a nonsense codon.

Langridge's study showed that an important fraction of the variants possessed significant β-galactosidase activity: sixty-three per cent had activity higher than ten per cent of that of the normal enzyme; fourteen per cent had retained more than twenty-five per cent of the activity. Another observation made by Langridge is that mutant enzymes are usually thermolabile: less stable than the ordinary enzyme, they rapidly lose their activity when heated for a few minutes at 57°C. About one half of mutant enzymes are much more thermolabile than the normal enzyme. Proteins show alternating 'insensitive' zones, in which an amino acid change has a small effect on stability, and 'sensitive' zones.

Today, sequences can be varied more systematically since at least five other amino acids besides serine – glutamine, tyrosine, leucine, lysine and tryptophan – can be inserted using nonsense suppressors. Jeffrey Miller's team prepared about three hundred variants of the gene for the lactose operon repressor. Only forty per cent of the mutant repressors had clearly altered properties. Non-polar residues were sensitive to substitution by polar ones, while polar residues could often be replaced by both polar and non-polar amino acids. Glass and Nene studied 570 mutants of the β polypeptide chain of *E. coli* RNA polymerase. About forty per cent were lethal and approximately two-thirds of the remaining 330 displayed various structural or functional defects.

The *ram* 1 mutants of *E. coli* make so many mistakes in translation that, out of every 100 β-galactosidase molecules which they synthesize, no two have the same sequence. This protein, it is true, is 1100 amino acids long. The levels of protein heterogeneity in the *ram* 1 mutants, evaluated with various criteria, are in the order of seventy per cent. The bacterium is viable even if it grows badly. These *ram* 1 mutants have defective ribosomes. Another way of muddling up translation is to introduce missense suppressor tRNAs which systematically insert a particular amino acid in response to another amino acid's codon – with 5–10 per cent efficiency. Bacteria are quite capable of growing despite the presence of the suppressor tRNA. But two missense suppressors would be too much. These results indicate that a point in protein space can be replaced by its immediate neighbourhood without seriously affecting cellular function: about a given activity there is always a dense population of points which correspond to the same activity.

In mountainous regions rivers flow straight over short sections, but their course is usually irregular with frequent changes of direction. In flat areas, rivers can form bends and sometimes intersect themselves, but the river usually has a very straight course. The Principle of Parsimony (Chapter 5) which states that molecular evolution takes the shortest possible route from one sequence to another, is in accordance with a fairly flat and gently rolling landscape. It is not surprising to find that sequence comparisons made in accordance with this principle give results fairly close to neutralist predictions: one gets back what one has put in.

Let us now come to the big question: what is the probability of finding an interesting enzymatic activity by choosing a point at random in protein space? Quite obviously, the interest of a given activity depends on its whole context, and this will be gone into more deeply in the chapters which follow, especially the last. Here I shall confine myself to evaluating the probability of fishing out a protein equivalent to a given protein, and I will use a novel argument which links the probability over the whole sequence space with local properties, discussed previously. Let us take as accepted the notion that if a protein mutates there is a probability p of obtaining a variant equivalent to the original protein. From what we have seen earlier, values of p between 0.3 and 0.4 are quite reasonable. If I start with a given protein and construct all the sequences which differ from it by one amino acid, I have by definition a probability p of finding the right activity again. I then make a great oversimplification. I suppose that if I change two amino acids, I will have a probability p^2 of having the correct activity, and so on: p^3 for three amino acids, p^n for the *n*th variants. By varying a larger and larger number of amino acids in this way, one ends by exploring the whole of sequence space. The total probability in this case is very simple: $P = [(1 + 19p)/20]^L$ where L is the number of amino acids in the sequence. For example, if p = 0.5 and $L = 100$, the probability is 10^{-28}. This number, obtained by a 'physical' argument based on experimental results, is much higher than everything proposed until now. There is a presupposition in this calculation: why do the probabilities vary as p^2, p^3, . . . p^n as we deviate from the initial sequence? Strictly, this calculation corresponds to the following (false) model.

Let us assume that there are N simple variants of the initial sequence having the right activity with no more than one variant per position. Then the double variants with enzymatic activity would be those obtained by combining two simple variations from among those which have retained the activity. Having revealed the weakness of the calculation, we can now see the way to refine it. We will have to introduce the probability of obtaining (or not obtaining) an 'active' double variation from the combination of two simple 'active' variations. The kind of knowledge which would allow these problems to be solved is presently being gained. It seems to me that a Principle of Equivalence or of local quasi-equivalence could be introduced: if two points in protein space represent equivalent sequences for their activity, there should be equal densities of points corresponding to that activity around each of the two points. In other words, p has the same value at every point in a continuum of sequences which correspond to the same activity.

The value calculated for the probability (10^{-28}), which is still very low, would be much higher if the calculation had been made for shorter sequences. Keeping p = 0.5, but with a sequence of thirty amino acids, P would be one chance in 230 million. We thus reach reasonable estimates. I have indicated why, in my opinion, ancient proteins were much shorter than contemporary ones (Chapter 11). If this idea is confirmed and it is shown that short proteins can have most of the activities of present proteins it will have to be proved that the value of p is within reasonable limits; in the order of 0.2–0.6, which seems physically plausible to me.

Once an interesting activity has been obtained with a short protein the size of the protein can be increased without too much difficulty, with a good chance of conserving the activity. A length of chain A–B–C–D–E on the outside of a protein globule could be replaced by a longer chain forming a protuberance:

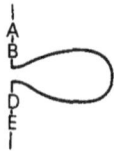

Such events, which certainly occurred in the evolution of nucleic acids (compare tyrosine tRNA to phenylalanine tRNA in Fig. 6), must also have played a part in protein evolution. Proteins can also elongate by joining globules together, without major modifications of their activities (Chapter 4).

Through the study of the properties of mutant enzymes, we reached the notion that the protein landscape was mildly undulating, as predicted by the neutralists. But now, there are two classes of arguments leading to a different perspective: neutrality may not be a general property of the protein landscape, but an acquired, evolved property. Conrad considers that mild amino acid substitutions are more manageable in protein evolution than drastic ones. If there is a genetic way of controlling the frequencies of the various amino acid substitutions, then organisms which change their proteins smoothly will be selected, and those prone to abrupt changes will be eliminated. Selection pushes the organisms towards neutral regions of the sequence space. I see two reasons why the observed smooth landscape around contemporary sequences may be an evolved artefact. First, neutrality means that there are many pathways permitting escape from a point in the sequence space. It then implies that there are many pathways leading to this point, making it an easily reachable one in evolution. Next, if the major steps in sequence refinement occurred under conditions of low accuracy in translation, then evolution was acting not upon the gene alone, but upon the gene plus its whole neighbourhood.

13 Acquisitive evolution

Throw into the fields a herbicide: a product of the chemist's imagination, newly synthesized in the laboratory, a substance never seen before in nature. Soon you will see bacteria growing which not only are not poisoned by the herbicide, but consume it as freely as if it were sugar. They have equipped themselves with all the enzymes necessary for digesting the new substrate. What has happened; what resources has the bacterium mobilized in response to the new situation? These are the kind of questions to which studies in acquisitive evolution, started only 12 years ago by Mortlock and Wood, are now providing answers which make the ideas of a whole generation of molecular evolutionists obsolete.

The bacterium which Mortlock studies, *Klebsiella aerogenes*, is equipped to attack a wide variety of substrates of the polyhedric alcohol family: glycerol, ribitol, arabitol, mannitol and sorbitol. Normally, it cannot grow on xylitol, a compound of the same series which is not found in nature. By imposing conditions on the bacteria which favours those which are nevertheless able to use xylitol a little, it has been possible to cause them to evolve into strains which can survive quite happily with xylitol as their sole source of carbon-containing food. Thanks to the work of Lin, we now know what resources were mobilized in response to the new situation. We shall also see how Lin resolved an even less trivial problem in the bacterium *E. Coli*.

If one causes *Klebsiella* bacteria to mutate and tries to culture them in the presence of xylitol as sole carbon source, nothing grows. This procedure is too brutal; the bacteria do not have the means to respond to such a radical change of food. A gentler procedure, which is closer to changes that occur in nature, is to give the bacteria some xylitol plus their normal food so that they grow. Those which, somehow or other, can make themselves a little extra from xylitol grow a little faster than the others, especially if the amount of ordinary nutrient is restricted. Thus the first stage in selection is the production of bacteria which can use a little xylitol, with low efficiency, but which cannot completely do without their usual nutrients. Biochemical analyses show that they have not acquired any novel catalytic activity. But they produce large quantities of an enzyme which attacks another compound of the xylitol family: ribitol. This enzyme, the first in the metabolic pathway for ribitol consumption, oxidizes it to ribulose. Like every enzyme, ribitol dehydro-genase does not have absolute specificity and can attack xylitol a little, in addition to its natural substrate, provided that it is present in the cell. The cell is not equipped to import xylitol but it enters by deception, since the cell's frontier posts do not have absolute specificity either. It has been known for a long time that enzymes may act on unnatural substrates: a large part of the enzymologists' knowledge is derived from model reactions using artificial coloured substrates, easier to characterize than their natural counterparts.

Normally synthesis of ribitol dehydrogenase occurs only when ribitol is present, otherwise it is repressed. Ribitol acts on a regulatory protein, a repressor, which normally blocks the ribitol dehydrogenase gene. In the presence of ribitol, the repressor detaches from the DNA and the gene is expressed. The first mutants, selected in cultures containing xylitol, are simply mutants in which the repressor is inactive so that ribitol dehydrogenase is made whether ribitol is present or not. Such a mutant is called 'constitutive' for the enzyme whose synthesis cannot be repressed.

Constitutive mutants then serve as the starting point for a new selection cycle and, at last, bacteria capable of growing on xylitol as sole carbon source are obtained. Three teams arrived at this result, using various mutagenic treatments and culture conditions. Hartley's team obtained bacteria which were over-producers of ribitol dehydrogenase: the gene had duplicated several times to give many copies so that the bacteria made enormous quantities of the enzyme. In this way ribitol dehydrogenase, which is not mutated itself, can constitute up to twenty per cent of total cell protein. But the bacteria do not go further along this pathway, for they get trapped by a second regulatory mechanism. According to Inderlied and Mortlock, the enzymes which follow ribitol dehydrogenase in the degradative pathway are regulated by the products of their reactions. If there is too much of them, the cell somehow deduces that the enzymes in the chain are too active and keeps them just ticking over. In certain cases, progress in the exploitation of xylitol is linked to changes in enzyme specificity. Lin's group observed a mutant growing on xylitol in which ribitol dehydrogenase had increased activity for xylitol (by a factor of 2.2 compared with the normal enzyme) while maintaining intact its capacity for attacking ribitol.

This mutant in turn is the starting point of a new selection cycle. This results in bacteria which can use xylitol as sole carbon source even when it is present in very low concentrations. Nothing is changed in the internal enzymatic equipment of the selected bacteria; the novelty is at the frontier which lets xylitol penetrate better than normally. A second enzyme, arabitol permease, makes its contribution. Its normal function is to facilitate the entry of arabitol into the cell. Once ingested, arabitol is treated rather like ribitol but by different enzymes. After a mutation in the gene for arabitol permease, the enzyme becomes capable of greatly facilitating entry of the new substrate. Overall then, xylitol use by the cell borrows steps from two parallel metabolic pathways, those of ribitol and arabitol. As shown in Fig. 17, xylitol dodges about, handled alternatively by enzymes of the two pathways.

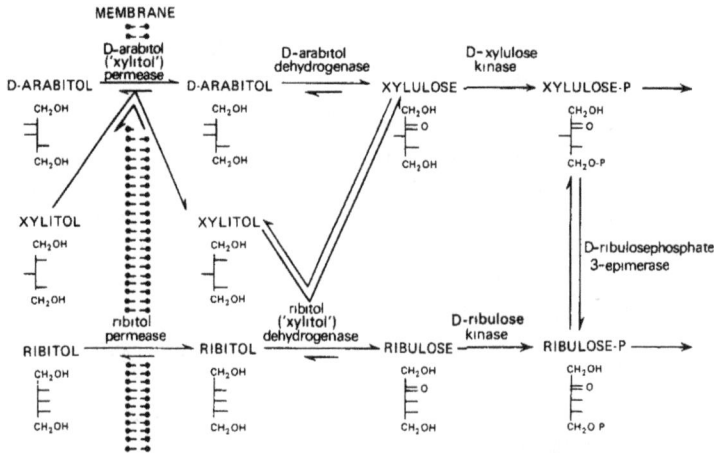

Figure 17   Utilization of a new substrate, xylitol, by taking advantage of pre-existing metabolic pathways. Xylitol creeps into the cell by the frontier posts for arabitol and is then attacked by the enzyme responsible for using ribitol. The product of this reaction is then taken in charge by an enzyme in the chain of arabitol conversion (according to Lin *et al.*, reference 12).

Experiments with other unusual substrates led to similar results. Patricia Clarke's team selected bacteria capable of using new substrates such as butyramide. At first, they obtained bacteria which produced large quantities of an enzyme – amidase – whose natural substrates are acetamide and propionamide. Although butyramide is not a natural substrate of the enzyme, it can be attacked by it with low efficiency. The constitutive mutants for amidase were then put through a second selection cycle by Clarke's team, giving *Pseudomonas aeruginosa* which were able to grow on butyramide as sole carbon source. This time the adaptation included a change in specificity: the activity of amidase towards the new substrate increased fifteen times.

Another strategy for producing novelty in bacteria consists of first suppressing a gene then attempting to select mutants which succeed in making up for the deficiency. Thus, the β-galactosidase gene in *E. coli* can be eliminated. When the bacteria are grown in the presence of lactose, the vanished enzyme's natural substrate, mutants are selected which make large amounts of another enzyme which is still poorly identified; let us call it ersatz. Normally, this enzyme can break down lactose, but is much less efficient than β-galactosidase. In a scenario which is beginning to be familiar to us, the constitutive mutants for ersatz pass through a new selection phase from which bacteria emerge that can use lactose as their sole carbon source. Hall, who conducted these studies, observes two types of mutant: those in which the specificity of ersatz is modified, and those which make much more ersatz than the usual constitutive mutants. In further studies, Hall succeeded in making the ersatz inducible again, the inducer being this time its novel substrate. He also obtained mutants of the ersatz that were able to hydrolyse lactulose, galactosylarabinose, or lactobionic acid.

Similar experiments by Kemper use an enzyme in the leucine synthesis pathway. One of the enzymes is composed of two polypeptide subunits. When the gene for one of them is eliminated the enzyme loses its activity completely. However, a mutation which has nothing to do with the genes for leucine synthesis allows the activity to be restored. Another enzyme in the cell, composed of two or more subunits, loses its cohesion as a result of a mutation and sometimes it lets one of its constitutive chains escape. This is then picked up by the forsaken subunits of the leucine enzyme and its activity is recovered. The same polypeptide chain is thus common to two different enzymes. This situation is perhaps more widespread than is usually believed. The elongation factor 'Tu' appears to be a polygamous protein.

Recently, Lin's team have succeeded in growing *E. coli* bacteria on propanediol as sole carbon source. This exploit is far from trivial, since propanediol is an excretion product of the bacteria, or waste, under certain conditions. The naïve thermodynamicist who sees the cell as an irreversible machine transforming fucose into propanediol would find it difficult to conceive that the same bacterium might start off with propanediol, function with the same efficiency, and moreover produce a little fucose. In fact, the bacterium is capable of growing according to two very different programmes. The first option is aerobic growth (in the presence of oxygen) and the second is anaerobic growth, in the absence of oxygen. Starting from either of these two extremes, the bacterium can go towards equilibrium from the point of view of the degrees of oxidation of its molecules. Prigogine's school, who present life as being fundamentally linked to processes occurring far from thermodynamic equilibrium, do not seem to be right.

Utilization of fucose passes by way of an intermediate, lactaldehyde (Fig. 18). During anaerobic growth lactaldehyde is reduced to propanediol, which is excreted. Let us call the enzyme responsible for the conversion oxidoreductase. This enzyme, present during anaerobic growth, is normally repressed when oxygen is present. The first mutation which takes the bacterium some of the way towards the use of propanediol is, once again, a regulatory mutation which maximizes oxidoreductase production whether oxygen is present or absent. Mutants which use propanediol more and more efficiently are selected from the first. The ability to consume the new substrate is increased further by regulatory modifications which affect other enzymes, both upstream and downstream from oxidoreductase in the pathway.

All experiments on acquisitive evolution have a common finding: two or three point mutations are sufficient for a bacterium to become capable of growth on a new substrate. In each case, it is found that enzymes do not have absolute specificity; one is always found which can perform the required reaction with very low efficiency. It is then sufficient to augment the level of the enzyme to meet the evolutionary challenge and to adjust the levels of

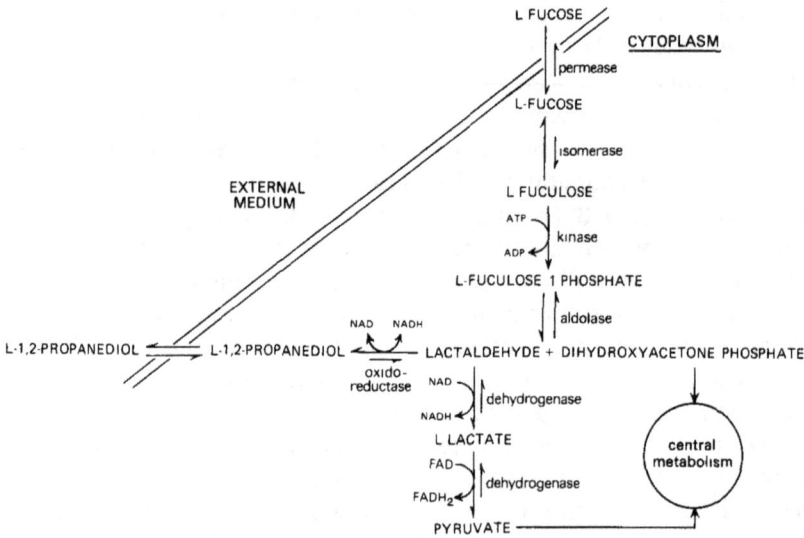

Figure 18   Inversion of a metabolic pathway. Some bacteria are capable of using fucose as sole carbon source in the absence of oxygen via the metabolic pathways indicated in the scheme. A degradation product, propanediol, is normally excreted by the cell. Mutants have been produced which can do the opposite and make fucose from propanediol, which they use as sole carbon source – although this does occur under different conditions (the presence of oxygen). (According to Lin *et al.*, ref. 12.)

other enzymes to form a well-balanced metabolic chain. Gene duplications allow production of enzymes with double tasks to be split up into production of enzymes specialized for each task. Changes of specificity are observed but they are much rarer than changes of level. These results accord well with the neutralist vision of the protein landscape containing great basins linked by large channels. When one speaks to the Darwinist about acquisitive evolution, he shrugs his shoulders and is hardly surprised. Mutant bacteria which can use a new substrate have a selective advantage over the others. So they will be fixed in the population. Such an answer bypasses the problem which interests the new generation of evolutionists, which is to define the range of possibilities: how and towards what is the bacterium likely to evolve; what evolutionary problems is it capable of resolving (through mutation and selection) and what are the evolutionary challenges which exceed its capacity to respond?

14 Molecular defences

At all levels of life, molecular mechanisms form barriers against novelty. These mechanisms certainly have other functions: to eliminate molecules foreign to the organism when they infiltrate, and to clean cells of their damaged molecules. The simplest bacteria have a set of maintenance enzymes: proteases and nucleases. This system is a sort of automatic kitchen robot with a unique program. At the top of the scale, in higher vertebrates, is found an extraordinary flexible system: immune defence. Its functioning, which is still not perfectly understood, acutely poses the problem of the relationship between mutation and selection, of generation of novelty at the molecular level and of exploration of sequence space. But let us start at the lowest level.

All proteins age. Amino acids alter spontaneously, especially glutamine and asparagine, which when hydrolysed lose ammonia and become glutamic and aspartic acids. The speeds of transformation of amino acids depends greatly on the environment of the residue in the protein. Certain proteins may change sequence in minutes while others take months. Nucleic acid bases also change spontaneously: cytosine loses its amino group and is converted to uracil. One of the functions of cellular proteases is presumably to degrade proteins which have changed sequence. The combined action of several proteases decomposes a protein into amino acids, which are used in new syntheses. Each protease has its own spectrum of activities: certain proteins resist it while it alters others without difficulty. Taken together, proteases constitute a filtering system incapable of distinguishing individual molecules. Rapid degradation of certain gene products is advantageous; for example, when they intervene at a precise moment in the cell cycle. For others it is important for the product to have a long life. To a certain extent then, it is the cell proteins which adjust to the protease system and not the reverse. When a gene mutates and produces an altered protein, this could well pass through the protease filter as if it were a normal protein; but quite often proteins produced after mutation are easily attacked by the proteases. In all cells which make many mistakes in translation, protease activity is intense. But novelty is also checked at the preceding level: messenger RNAs produced by mutant genes are often rapidly destroyed by nucleases. The nuclease and protease filter attenuates novelty but it does not eliminate it completely. If the novelty is particularly advantageous, it is not impossible for the filter to change its specificity so as to allow the new protein to be expressed. Since a mutant protein, which still strongly resembles a normal protein, is likely to be degraded, a protein which is completely foreign to the cell should run an even greater risk, although this is not certain. Certain nucleases in the cell probably constitute a line of defence against invasions by foreign nucleic acids; and in good arms-race logic, some phages carry proteins which knock out bacterial proteases.

In higher organisms, immune defence accomplishes a much more elaborate task. It produces antibodies which help to eliminate the widest range of substances – sugars, proteins, nucleic acids and barbarian chemical compounds conceived in the laboratory – with the exception of molecules normally present in the organism. The repertoire of responses is extremely varied, covering almost everything which is foreign to the organism. But the notion of foreign molecules is far from trivial in higher organisms. Every individual is an absolutely new combination of genes provided by each parent. If a gene has two alleles A and a, and a child a/a is born from a father A/a and a mother A/a, the child's antibodies will recognize the product of gene A as foreign even though it is present in both its father and mother. As Woese says, the antibody repertoire is custom made, unlike the protease and nuclease system. Another peculiarity of immune defence is its perfectibility. Faced with a first invasion of foreign molecules, the organism mobilizes relatively inactive antibodies. Over the next few days, its response improves and reaches a plateau. When a second invasion strikes the organism is capable of immediately synthesizing highly active antibodies. There are several ideas which help to clear the mystery and provide plausible molecular blueprints for the properties of immune defence.

The first important notion is that the immune system permanently possesses a great variety of antibodies and that a foreign molecule which gets into the organism binds, through random encounters, with the antibodies for which it has the greatest affinity. This association sets off a process which leads to increased production of the same antibodies. The commonest form of this idea (which is certainly not correct in every detail) is that each particular antibody is synthesized by a cell or family of cells – the lymphocytes – which advertise their product by carrying antibodies on their surfaces. Let us assume that every antibody can bind two antigen molecules (which is certain) and that each antigen binds to two antibody molecules (the official doctrine states that this happens provided that the two antibody molecules are of different structure, since they touch different parts of the antigen molecule). If the antigen is present in large amounts and the density of antibodies on the cell surface is high (and they have the right affinities for each other), a crust is formed in which each antigen is linked to two antibodies and each antibody is linked to two antigens. Normally, the formation of the crust induces the lymphocyte to multiply. Thus the foreign molecule encourages proliferation of the cells which produce the antibodies to which it can bind. After this, the large numbers of antibodies produced help to eliminate the molecule by a process which is rather well known but which I shall not describe here. A second idea, less well established than the first explains how the immune system does not attack the organisms's own molecules. When the crust forms at an early phase of maturation of the immune system, the lymphocite does not proliferate but is, instead, paralyzed and leaves little descent. 'Foreign' for the immune system would therefore be synonymous with 'absent at birth'.

Some points remain to be examined before dealing with the great difficulty which this account has glossed over. First, a large variety of immunological effects exists: paralysis, tolerance, suppression of tolerance and several types of cell are all affected by the immune response, or rather responses. There are many models which attempt to explain this variety in terms of quantitative effects which modulate or amplify the basic phenomena. The process by which the cells move from production of an already active class of antibodies to a much more effective class of antibodies poses no major problem. Without changing the specificity of the site on the antibody to which the antigen binds, it is possible, in principle, to make the antibody much more effective by making it generally more apt to aggregate into large networks. I am now going to present immunology in a more traditional manner, which will ensure that we no longer understand anything.

Since the antibody can bind efficiently and specifically to the antigen, it must possess a very special site with a structure complementary to that of the antigen; the antibody would be like a lock which can be opened by only a single key. Therefore each antibody must have a unique structure, adapted to that of its antigen. Now an individual has the capacity to respond to 100 000 or even a million different antigens. So it possesses at least an equal number of antibodies and genes which code for the antibodies. It would then be necessary for most of the DNA to be devoted exclusively to coding for antibodies. This seems to be out of the question. The difficulty can be avoided by assuming that initially the DNA codes for only a small number of antibodies and that later the cells of the immune system mutate: they would play 'roulette' with certain DNA segments and thus generate an immense diversity of antibodies from a few archetypes. This theory predicts that antibody sequences are related and this is what is found: antibodies form a multidimensional family like the haemoglobins. Like them, antibodies are double dimers. They contain a heavy chain (H) in two copies and a light chain (L) also in two copies. A heavy and a light chain associate to form an H–L dimer and the antibody is formed from two (sometimes more) H–L dimers. Just as in man one finds haemoglobins $\alpha_2\beta_2$, $\alpha_2\gamma_2$, $\alpha_2\delta_2$, etc., one finds several classes of antibody which are distinguished in particular by the properties of their heavy chains and which intervene at different stages in the immune response. There are remarkable sequence homologies between L chains of the same class from the same species and there is no doubt that L chains descend from the same common ancestor. Variation from one L chain to another is localized particularly in 'hypervariable' zones which are, as X-ray studies have shown, just those parts which make contact with the antigen. Just as in the haemoglobins, kinship seems even more marked in the three-dimensional structure than at the sequence level. So antibody genes duplicate and mutate.

But then we encounter a problem; because if, as orthodoxy would have it, each individual starts off with a small number of genes which then mutate the novelty engendered by mutation will be regarded as foreign. Newly invented antibodies would have to be eliminated by other antibodies, which in their turn, etc. The immune system would destroy itself. To resolve this difficulty, Jerne's Network Theory, currently in vogue, proposes that at an early stage of embryonic development a sort of war of antibodies takes place from which later emerges the family of cells which will ensure the defence of the individual.

The properties attributed to antibodies in the mutational theory certainly give the evolutionist something to dream about. Here we have a class of proteins which by accumulating a few mutations become capable of binding, with remarkable specificity, tens or hundreds of thousands of different substrates. Despite this, they show extraordinary homogeneity in their three-dimensional structure and, besides binding the substrate, they participate in a whole series of interactions with various components of the immune system (lymphocyte membranes and complement). Until further notice, antibodies fix substrates but do not catalyse any chemical reactions. Extraordinary mutations fashion the active site at will, giving it the shape required for binding the substrate but (nearly) always leaving it chemically impotent! We saw in the previous chapter how difficult it is to produce by mutation a protein with altered specificity. It is necessary to breed thousands of millions of bacteria to increase an enzyme's activity by a factor of five or six with a substrate which it can already bind; antibody genes are now being asked to produce, after a few mutations, proteins a 1000 or 10 000 times more active with any substrate whatever, given in advance. In this conflict between traditional immunology and traditional evolutionism, I opt for the concepts of the second discipline. We must now see why we let ourselves get into this impasse.

It is clear, in my opinion, that we must re-examine the initial description of antibody specificity. No, antibodies are not as specific or exclusive as was claimed. They are no more specific than activating enzymes, which we shall speak about in the next chapter. With a thousand heavy chains and a thousand light chains a million H–L dimers can be made, so a million genes are not needed to code for all the antibodies. Talmage and especially Inman defend the 'combinatorial' theory of antibody specificity. One can go further. The L chain is synthesized from several portions of different genes and complete fusion of the pieces occurs late in development. With thirty 'left-half' genes and thirty 'right-half' genes, 900 complete L chain genes can be made. So an immense potential repertoire of antibodies can be coded for by a small number of genes.

Take a foreign molecule and a preparation of antibodies. Eliminate those antibodies that have a strong affinity for the molecule. Now, dissociate the remaining inactive antibodies and reassociate them at random, thus forming antibodies with new combinations of light and heavy chains. The mixture now contains antibodies with a measurable affinity for the foreign molecule, proving the combinatorial nature of antibody specificity. Actually, this is a thought experiment that none of the immunologists I know finds tortuous enough to deserve being carried out.

To fight any given antigen, the organism produces an enormous variety of antibodies. When one of the corresponding lymphocytes is isolated and cultured in the laboratory, it gives rise to a colony that produces a unique antibody molecule. When such 'pure' antibodies are examined, they are found to be able to bind, with an appreciable efficiency, molecules other than the target antigen. Pure antibodies make mistakes. But in a natural heterogeneous population of antibodies, each component makes a different mistake so that errors are less visible, and the population as a whole is much more specific than any of its individuals.

It is easy to imagine mechanisms by which the response improves, in the combinatorial theory, without bringing in mutations. For example, if a cell makes the antibody H_1–L_1 and this binds an antigen well, the antibody will be able to cause a special class of cells to make antibodies with the same light chain L_1 and any heavy chain. Among all the H_n–L_1 products, there will be some which will be more effective then H_1–L_1. Then the heavy chain can be conserved and the light one changed. I do not exclude mutations at the embryo stage provided that the amount of variation is strictly programmed at the beginning. Mutations here would not be a source of novelty but the corollary of a compact coding, the means of exploring a repertoire potentially coded for at the beginning. In this view the antibody repertoire is 'closed': the possibilities are well defined at the start and the individual expresses all of them, thus exposing the whole of its immune mechanism to natural selection. These theories are opposed by those who assume an 'open' repertoire in which each individual produces a great variety of antibodies which its parents would never have possessed. For me, the demarcation is not between theories with and without mutation (in classical terminology, somatic or germinal theories) but between closed and open repertoires. The first are compatible with classical evolutionism, the second call for serious adjustments, which makes them interesting. We know of two great families of haemoglobins: that of the principal chains found in living species, which are stable and effective, having been selected during evolution; and the family of abnormal haemoglobins, which reflect sudden variations in the genes. By certain features, notably thermal stability, an open repertoire should be related to the abnormal haemoglobins or to the β-galactosidase variants presented in Chapter 12.

If I am wrong about the closed nature of the antibody repertoire and immune defence really does have the creative properties which some people suppose it to have, then so much the better. By slightly modifying the antibody-producing system, which is so well run-in, there would be no difficulty in developing a universal system of enzyme production, performing at will the required reactions chosen in advance. The theory on which I am casting doubt for the moment could well turn out to be that of progress.

Some micro-organisms have developed ingenious counter-strategies to deceive the immune defences of higher organisms. Thus, African Trypanosomes that invade the bloodstream of mammals are all-covered on their surface by one protein which is the natural target of antibodies. But the species owns more than a 100 genes coding for different surface proteins. Each individual contains the complete repertoire of genes for the potential surface proteins, but expresses only one of them thanks to a mechanism involving the duplication of one of the basic genes and the transposition of the replica to a site in the genome where it can be transcribed. Now, in a host, the population of parasites will contain, say, ninety-five per cent of cells with proteins of a certain type at their surface. The immune system will develop defences against those cells and eliminate them. But then, the remaining five per cent will be able to grow, and there will be another major surface protein for the next round of hide and seek.

15 **Molecular crosses**

When a horse mates with an ass, the fruit of their love has all the vigour and intelligence of its parents but it is sterile. This cross-breeding produces a viable egg which passes through all stages of embryonic development to result in an animal which is whole from all points of view, with four legs of equal length and a brain in good working order. This tells us much about the logic of genetic programmes of development. Those who find pleasure in measuring genetic distances between species will conclude that the ass and the horse are very close relatives. Here we see outlined a new approach to the measurement of genetic distances between species: a female is fertilized by a male of another species and the number of developmental steps which the hybrid is capable of surmounting is determined. This kind of study is practised successfully in plants. Plant hybridization sometimes leads to new, viable species which are not simply laboratory curiosities. The idea of comparing different species, not by measuring their degree of similarity but by making them interact, is transposable to the molecular level. Is a bacterial messenger RNA correctly translated in the test tube by human ribosomes? Is yeast RNA polymerase capable of transcribing DNA from *Bacillus subtilis*, starting and stopping at the right places? Molecular biology has accumulated a mass of information of this kind, the most interesting of which goes against intuitive thinking: the greater the genetic distance between two species, the better their molecules interact!

Let us take the activating enzyme for phenylalanine in *E. coli*. It cannot charge the tyrosine-specific tRNA from the same organism, but has no difficulty in charging the tyrosine tRNA from yeast.

Studies of this type were undertaken to refine structural comparisons: if an enzyme charges both tRNA A and tRNA B then A and B are similar, at least in that portion of the molecule which the enzyme recognizes. tRNA was thought of as a mosaic of 'specific sites', one for attachment to the ribosome, another for recognition by the activating enzyme, etc. It was rather like the gourmet's portrait of a bullock which adorns so many kitchen walls, in which dotted lines separate the different parts of the anatomy: sirloin for the ribosome, undercut for the activating enzymes, the skin reserved for the elongation factors. Each activating enzyme would thus interact with a particular portion of tRNA and would charge it only when this portion contained the sequence specifically recognized by the enzyme. Biochemists used to believe that, in the cell, the activating enzyme recognized its own tRNA or tRNAs and ignored the rest. In fact it interacts to a greater or lesser extent with a third of the tRNAs in the cell and is capable of charging more than one. But, when it has charged a 'bad' tRNA, it immediately corrects the mistake, rather like a DNA polymerase. It is equipped with a security device against the tRNAs which it ought not to charge.

Studies aimed at working out rules for recognition of tRNAs by activating enzymes have led nowhere. It is not possible to pick out a segment of a sequence and say that this is what the activating enzyme recognizes. Yarus even points out that resemblances are to be found in the group of tRNAs which are not charged by the enzyme – the excluded ones. Specific recognition appears to result from a general mode of interactions onto which are imposed specific measures against undesirable substrates. The perfected defences which the activating enzyme has against undesirable tRNAs of its own cell become ineffective with foreign tRNAs which it charges because there is nothing to tell it to exclude them. Table 3 shows the results of

Table 3 Charging of the tRNAs of one species by the activating enzymes of another species. Results are given as percentages*.

Source of activating enzyme for valine:	B. stearothermophilus		Yeast
Source of tRNA:	Yeast	E. coli	B. stearothermophilus
Valine	78	63	74
Phenylalanine	62	1	17
Isoleucine	12	1	30
Methionine	9	2	68
Tyrosine	8	4	9
Tryptophan	6	5	0
Alanine	5	3	61
Lysine	5	1	12
Arginine	4	0	2
Threonine	3	0	23
Glycine	3	0	7
Histidine	3	1	9
Leucine	2	0	13
Glutamine	2	0	6
Proline	1	0	43
Serine	1	0	9
Glutamic acid	0	0	6
Asparagine	0	0	3
Aspartic acid	0	0	3

* Cross-interactions are stronger between distantly related species than between closely related species (central column). See reference 124 for experimental conditions.

cross-interactions between tRNAs and activating enzymes from three species: yeast, a bacillus and *E. coli*. These are the kind of data which one would like to have to define the properties of sequence space. At each point in sequence space which represented a protein, we assigned a height which corresponded to the activity of the protein. But it is not legitimate to think of the cell as the sum of individual activities. Here we begin to see the link between different molecules in the same cell. When a tRNA sequence evolves, how is the tRNA charged by each of the cell's activating enzymes? Experiments on charging between tRNAs and activating enzymes from different species give a partial answer to this question. If we isolate the twenty activating enzymes and the forty tRNAs from several species and study for each pair of species the $20 \times 40 = 800$ cross-interactions, we should obtain extremely rich information and be able to establish from it interspecies distances no longer based on structural criteria but which would be 'functional distances'.

In Chapter 3 I introduced the organelles, present in the cells of higher organisms, which are themselves sorts of miniature, largely autonomous cells. Thus, there are in the cell a 'central' translation apparatus which functions in the cytoplasm, a translation apparatus in mitochondria and – in plant cells – a third one in the chloroplasts. Each of these apparatuses has quite clear-cut characteristics which distinguish it from the other two. Some people consider that long ago the primitive ancestors of organelles were complete beings, living as parasites on the ancestors of eukaryotic cells (and incidentally to the latter's advantage) until they were swallowed by the host cell. Even though the ribosomes of mitochondria and chloroplasts are very different from those of the cytoplasm, all their proteins are synthesized by the central apparatus. If the genes for mitochondrial ribosomal proteins were initially part of mitochondrial DNA, they must have been transferred later to the central DNA. But there are limits to this transfer of responsibilities, since the central apparatus cannot simply replace mitochondrial ribosomal proteins with its own ribosomal proteins. Nevertheless, it can, to a certain extent, supply tRNAs usable in the mitochondria or chloroplasts. The cells of higher organisms would thus appear to be the fruit of a very special cross in which fusion is incomplete, each ancestor perpetuating some of its own traits in the offspring. Let us extend the molecular analysis of this situation further.

Initially we have two translation apparatuses functioning in parallel, one in the host-cell's cytoplasm and the other in the ancestor of the organelle. If there is communication between the organelle and the host-cell's cytoplasm and the components of the translation apparatuses (tRNAs and others) differ from one compartment to the other, errors will be produced, just as when tRNAs and activating enzymes from two different species are mixed. Worse yet, some codons may correspond to different amino acids in the two apparatuses. The later evolution of the two apparatuses will obey two contradictory tendencies: to reduce the number of genes which do the same job and to maintain precise translation in the two apparatuses. Can we deduce from the foregoing any rules about the characteristics of translation components in the two compartments? Here are some of the early guesses I made, to solve this problem submitted to me by Orgel.

Initially, the organelle membrane is slightly permeable; communication with cytoplasm is possible but autonomy is preserved. How can communication be maintained while avoiding errors due to partial mixing between the two translation apparatuses? Two readily applicable solutions may be glimpsed here. Activating enzymes apt to make mistakes would be confined to their cellular compartment following mutations which would give them an affinity for the membrane of their compartment or for other proteins of the same apparatus. Gross aggregates would have difficulty in crossing the organelle envelope. Enzymes could 'stamp' molecules synthesized in one compartment by adding on methyl groups, phosphate groups or others which would make them suspect beyond their own borders where they would be degraded by cleaning enzymes. Or one can envisage a slow evolution of each component of one of the apparatuses to minimize interactions with components of the other. Cross-relations between tRNAs and elongation factors or activating enzymes would thus be quite different from those observed when one compares translation apparatuses of two species, whether they are closely or distantly related. As the translation apparatuses adjust to one another, it becomes possible to suppress parts which are doubly employed. It would be difficult to get rid of the activating enzymes of serine, leucine and arginine which have the delicate task of charging four or five different tRNAs. Besides, an organelle tRNA can be replaced by one from the central apparatus on the condition that all codons continue to be read. It will be easier to replace a tRNA which reads the codons C_1 and C_2 by one which reads C_1, C_2 and C_3 than the reverse. Therefore we can foresee that in the organelle there will be a tendency for tRNAs which read a single codon to disappear. The 'coverage' of the codons must be particularly large in organelles. Activating enzymes do form aggregates in the cytoplasm and the codon coverage is large indeed in organelles. Unforeseen was the fact that the genetic codes would differ. Perhaps they were different at the start. But it is also tempting to speculate that in some cases the error-level in the translation of some codons was so high that a change in the meaning of these codons could not have made the situation much worse, and was therefore tolerable.

The degree of structural resemblance between two proteins of unknown sequence may be evaluated by drawing on the paradoxical properties of the immune response. Let there be a bacterium which produces an enzyme E which one recognizes by its activity. One purifies E then administers it to rabbits, which proceed to make anti-E antibodies. Let us suppose that a mutation changes E into an inactive protein E'. How is E' to be detected? Anti-E antibodies are added to cellular juice containing proteins which is extracted from the bacteria. These act on E'; less well than on E, but enough for E' to be precipitated and so purified. The technique relies on the fact that the discriminating power of antibodies is not absolute, that E′ is to a certain extent confused with E. Let us now take a haemoglobin of species A, which we shall call HbA, and make an anti-HbA antibody. Next we react these antibodies with a whole series of haemoglobins from different species and measure quantitatively the effectiveness of each antibody–haemoglobin interaction. The closer a haemoglobin sequence is to that of HbA the better it reacts with the antibody raised against it. In this way we gain a measure of 'immunological distance' between sequences, which moreover correlates well with that based on amino acid differences. This methodology was successfully developed by Allan C. Wilson and applied notably to determining the phylogeny of the primates, where it compensated for the uncertainties of palaeontology.

Immune defence distinguishes between self and non-self; a molecule should provoke a livelier response the more foreign it is to the organism producing antibodies. On this basis, the distance between two species A and B could be calibrated by measuring the avidity of antibodies produced by A for B's antigens. In Wilson's work, avidity varied *inversely* with distance. But in this case the antibodies were provided by a 'third party', an animal sufficiently distant from the sources of the antigens for it to be considered as impartial with regard to them. There still exists a detour by which antibodies allow relationships to be evaluated. Closely related species are susceptible to infection by the same bacteria or viruses or colonization by the same parasites. Let us suppose that species A is commonly infected by a virus V, which can be isolated. Species A individuals, infected by the virus, contain anti-V antibodies. Is species B also attacked by virus V? To find out one takes antibodies from B individuals; if they precipitate V it shows that B has been infected by a virus similar to V.

Many enzymes contain two or more polypeptide chains. In the allosteric enzymes in particular, when a regulatory signal brings one chain into service the others are set into action in unison. Enzymes which have homologous functions in different species often, but not always, have the same mode of assembly in subunits. Thus all haemoglobins, apart from certain pathological cases, contain two types of chain with two copies of each (*see* Chpater 3). Can a dog's α chains associate with β chains of his master to form a functional α_2 dog β_2 man haemoglobin? The answer is yes, at least for binding oxygen. This is true for hybrids between man and dog, mouse and ass, etc. But certain fine properties of the reaction (allosteric response) are generally lost in the hybrids. There are infinite variations on this theme. The problem can also be turned around.

Normally in the cell, enzymes have well-defined subunit composition; it is difficult to imagine a dehydrogenase wandering about with a chain borrowed from haemoglobin. Protein subunits recognize one another. If one denatures a mixture of proteins, dissociating all the multimers into isolated chains, then re-establish conditions which favour association, one recovers the initial enzymes intact without forming monsters. Cook and Koshland found this in every case they studied. Experiments which were performed some time ago deserve to be repeated, since results produced in a given period tend to agree perfectly with the concepts of the time. When the concepts change, one perceives that the authors of old experiments contrived to choose the experimental set-up which least risked producing results contrary to the dominant concepts. I described in Chapter 13 Kemper's experiments which provided evidence for the formation of molecular monsters *in vivo*.

Hybrids between DNA or RNA chains from different organisms are studied much more systematically than protein hybrids. When a solution of DNA is heated, the two chains separate at a temperature which depends on the base composition of the complementary strands and also on the DNA concentration. Chains rich in adenine and thymine dissociate around 50°C, whereas chains rich in cytosine and guanine do so at 80–90°C or above. Let us take two closely related species and form a hybrid between a DNA chain from one of them and the chain complementary to its homologue in the other species. If the homologous sequences contain differences, the two chains in the hybrid will not be exactly complementary and they will dissociate at a lower temperature than the two double-chains from which they originate. The temperature difference provides a measure of the dissimilarity between the DNAs of the two species. Nucleic acid hybridization techniques are used a great deal in molecular biology. They were exploited by Britten, Davidson and Kohne to evaluate sequence similarities in nucleic acids.

As the method relied on rather doubtful assumptions, it proved to be extremely powerful for locating genetic segments that were not sought for, leading to unexpected important discoveries on the structure of the genome. I outline here the image of the genome which emerges from the hybridizations and sequencing work.

The precise order of the base pairs in a chromosome is not rigidly defined. there are mobile sequence elements that can be excised from one place and inserted at one of a few other places. Most often, but not always, such transpositions are accompanied by duplications of the mobile element. In a few cases, the mobile sequences contain structural genes, or play a regulatory role, helping to switch on or off a remote gene. But usually, their sequences, repeated with variations several hundred or several hundred thousand times, do not make any sense. Are they, as Dawkins would call them, 'selfish genes' that have managed to multiply within the chromosome, without benefit to the species?

In higher organisms, the genes for ribosomal RNAs are repeated head to tail thousands of times, and are located at strategic positions in the chromosome. They vary very little from one species to the other and have been used to compare the most distantly related species (Chapter 3, p. 25). The genes for histones also belong to long repetitive zones and have a low mutation rate.

Most of the protein sequences are encoded by a unique gene. It has been suggested that the introns – the non-coding sequences which in higher organisms interrupt the genes – were formerly movable elements that anchored within the genes, then degenerated and lost their mobility. Later on, some introns might have become, as suggested by Slonimski in the case of yeast mitochondria, genes in brackets, coding for a protein of their own, the function of which would be to control the splicing of the enclosing gene. Sometimes, a messenger RNA is copied back as a piece of DNA and integrated into the chromosome, becoming a 'pseudogene', recognizable from a functional gene. They resemble pieces of messenger RNAs that had been incompletely processed, with some introns spliced out, and an added tail of Poly(A). Their codon composition is unconstrained, and the sequences show high variability in-between species.

The single-copy genes generally alternate, in higher organisms, with highly repetitive, fast-evolving sequences. A remarkable fact is that whilst evolving they retain their repetitive nature, which suggests that they are often regenerated from a few copies. In most studies on acquisitive evolution described in Chapter 13, it appears that molecular evolution proceeds initially by mutations in regulatory genes. Since short repetitive sequences flank structural genes, it is tempting to attribute a regulatory role to them. Their great variability would appear to approach that inferred for bacterial regulatory genes. Conversely, the low variability of long repetitive sequences fits in well with the idea that these zones contain genes known for their stability during evolution (e.g. ribosomal RNAs and histones).

16 The great error loop

In population genetics, the actors are individuals possessing well-defined characters. From one generation to the next, characters remain but their proportions in the population change. Under prebiotic conditions on the other hand, a DNA molecule which replicated would be highly unlikely to leave descendants identical to itself and the characteristic of molecular populations would change at each generation. How can we argue in these fuzzy, changing situations? I do not have a complete answer to this ambitious question but I present here arguments which allow us to outline evolution in the transition period between the prebiotic and Darwinian regimes.

Cyberneticians feel completely at ease when they discuss cellular processes. They see the cell in the image of a man-made machine. When a button is pressed, the machine starts up and makes the required product. Except unlike a mechanism made of levers and gears, the cellular bag contains molecules which move in all directions, hit one another and deform. Molecules are constantly synthesized and destroyed. On a superficial level the cyberneticians are right; the cell functions like a well-oiled machine. We have to bridge the gap between the regular functioning of the cell and its molecular description with all its hazards and complexity. A good appreciation of this link is, in my opinion, indispensable for attacking the problem of the origin of life.

The following argument, made by Orgel and presented in its initial simplicity, leads us to the heart of the matter. Let us assume that an error is made during the biosynthesis of a protein and that this protein (an activating enzyme, for example) belongs to the translation apparatus and is partly responsible, for precision in decoding. Being faulty, it will cause new errors, some of which will affect proteins of the same type, and so on. Errors will be propagated, inexorably leading, after several generations, to the death of all the cell's descendants. If this description were absolutely pertinent, all life would extinguish itself; evolution would be impossible. Several adjustments must be applied to it.

Cell mass increases at a certain speed and errors are propagated at their own rate. Even if each error gives rise to new errors, this propagation will not be catastrophic as long as it does not take on the general rate of increase of cell mass. In addition, there could be compensatory effects: some errors could lead to more-precise proteins than before (moreover it appears that in present-day cells it is easier to mutate towards precise ribosomes or polymerases than the reverse; this may be a modern security device). Mutations which slowed down an enzyme's activity could in so doing increase its precision. Lastly, let us note that propagation of errors can happen in two ways. When a tRNA becomes a missense suppressor after a base change in its anticodon, it will systematically read the codons of an amino acid other than its own. However, the most common case is that of an enzyme which makes mistakes more often than the standard enzyme, the type of error remaining the same. At every moment in the cell there can be some copies of every protein which deviate from the prototype. In terms of sequence space, when the level of precision is lowered there is a deviation away from the hole occupied by the prototype; either by straight but long routes when the alteration is of the first type; or to an enlarged region around the initial hole in the second case.

With Hoffmann's model we address the first profound treatment of error catastrophe. His attempt has the merit of clearing the ground. He retains three classes of compounds between which everything takes place: amino acids, tRNAs and activating enzymes. Let $a_1, a_2, \ldots a_n$ be the amino acids which occur in proteins (here n is not necessarily equal to 20 since we want to discuss possible former states of the code). Let $T_1, T_2, \ldots T_n$ be the tRNAs which correspond to the amino acids. For simplicity, there is only one tRNA species per amino acid. Lastly, let $E_1, E_2, \ldots E_n$ be the activating enzymes. Normally, enzyme E_i attaches amino acids, a_i to tRNA T_i with a speed V. But in this model, it can also attach amino acid a_i to tRNA T_j (i being different from j) with a slower speed, W. The tRNA T_j will thus be charged either with the amino acid a_j when it interacts with its activating enzyme E_j or to a smaller extent (W is much smaller than V) with any of the other amino acids. The codons which correspond to the amino acid a_j are read by the tRNA T_j; for simplicity we do not take into account errors in codon–anticodon recognition. When tRNA T_j carries an amino acid other than a_j it is the source of a translation error.

Given well-defined values for the precision of each activating enzyme, it will be possible to deduce the proportion of incorrectly charged tRNAs and from this the rate of errors in protein synthesis. This is phase 1 of the model. In phase 2 one tries to evaluate the properties of newly synthesized proteins considering only activating enzymes. Given the error rate determined in phase 1 of the calculation, one can find the proportion of proteins synthesized which correspond exactly to the gene and how many differ from the prototype in only one or two positions. Then begins the really delicate part of the model. Among the altered enzymes, Hoffmann distinguishes two classes: those which have become totally inactive as if they no longer existed, and those which have lost their specificity for tRNA and charge all tRNAs with the same speed, W. To make the model more precise, Hoffmann assumes that the sequences of activating enzymes contain a defined number of vital sites at which an amino acid change completely inactivates the protein. At non-vital sites, such changes only make the protein lose its specificity.

Two opposing tendencies are at work. On the one hand there is a minimum level of errors, since even enzymes conforming to the prototype make them; the errors propagate and their effects are felt at the level from which they originated, hence a tendency to accumulate errors. On the other hand, the model supposes that faulty products are either completely or partially inactive (W is slower than V). They therefore have less influence on protein synthesis than correct products. In a way they eliminate themselves, hence a tendency to regression of errors. It seems intuitively correct to suppose that for certain suitable choices of parameters of the model (L, V/W) the two tendencies will balance one another and lead to a steady state. Let us now assume that we introduce a large number of errors into the system (or that there is a gross natural fluctuation of errors). If the injection of errors is too massive there is catastrophic accumulation; with a small fluctuation of errors a return to the stationary state will be possible. Every detailed treatment of error catastophe will almost certainly lead to the same qualitative predictions: possibility of the existence of one or several steady states in which errors are propagated at the same rate as cellular mass, return to the steady state for small fluctuations and change of state or catastrophe for gross fluctuations. Let us now change vocabulary. Let us say that we start from a situation in which the error rate is larger than in the steady state, without being at a catastrophic level. The system evolves, approaching a steady state. So it evolves from an initial low-precision state to one of greater reliability. Hoffmann's article is entitled *On the Origin of the Genetic Code and the Stability of the Translation Apparatus*. And it really is about the origin of the code: the appearance of ordered states from less-ordered ones. Such treatments are very important. A theory of the origin of the code postulates a succession of states through which translation apparatuses might have passed. For each of these states the question can be posed: was it sufficiently coherent to have been able to exist? Did its logic allow it to maintain errors at a stationary level or was error catastrophe inevitable? This kind of approach provides a theoretical sieve for evaluating models of the origin of the genetic code and it is the only sieve available. Intuitively, the more complex a system is, the more it requires precision for its maintenance. Statements like: 'such and such a translation apparatus was established because it had selective advantage over the others' are mere verbiage. Before generously attributing selective advantages to it, one should see if the apparatus can function coherently.

The choice of physical parameters in Hoffmann's model can be contested, in particular the notion that proteins which make many mistakes are less active than the others. In many cases loss of precision goes hand in hand with an increase in activity. In Hoffmann's model the effect of a sequence alteration is always severe. One is in an ultraselectionist universe: the correct protein's hole in protein space is hardly accessible and it is hard to understand how evolution was able to discover it. But the fundamental vice in Hoffmann's model (and in other models which it has inspired) is that error is defined as *deviation from an archetype, given at the outset*. We find again in Hoffmann, a former student of Eigen, the same finalist reasoning as in his teacher. An activating enzyme has less chance of being active in this model the further removed it is from the archetype, which is the ultimate state to be achieved. The low activity of enzymes containing errors is really the equivalent of natural selection, the sanction which hits bad molecules. As with Eigen and Schuster, selection acts by comparing the state at a given instant with the archetype towards which it is directed. Another fundamental reproach is that evolution of accuracy is due mostly to change in the discriminatory power of enzymes. In Hoffmann's argument there are good enzymes and bad and nothing in his approach allows modifications of enzyme specificity to be dealt with. In fact this is not out of reach.

In the treatment which I now present there is no longer any archetype. We can even erase the word error from the vocabulary and speak only of discrimination between various substrates. Like Hoffmann, I begin by focusing my attention on the system of tRNAs and activating enzymes and I simplify matters by limiting myself to n amino acids, n tRNAs and n activating enzymes. Let us call every RNA molecule 75–90 nucleotides long a potential tRNA and let all the potential tRNAs be represented by points in a plane: the 'tRNA space', like protein space. In what follows I suggest that the non-mathematical reader jumps as required over the mathematical symbols but nevertheless tries to follow the essential argument. Let us measure for each activating enzyme the efficiency, E_k, with which it charges each potential tRNA. The results can be represented in the form of a histogram (Fig. 19) in which the abscissa carries successive arbitrary values of the charging efficiency: $e_1, e_2, \ldots e_n$ and the ordinate has the number of potential tRNAs which are charged with an efficiency between e_1 and e_2, e_2 and e_3, etc. The histogram can be put into a 'continuous' form, more manageable mathematically. This time, we plot on the ordinate, for each efficiency value, a 'probability density' of potential tRNAs (*see* Fig. 19). Let us now put the real tRNAs of the cell in this diagram. For simplicity, I provisionally assume that there are strictly n tRNA species and n species of activating enzyme without minor variants. We can attribute an efficiency of charge e (T_i, E_j) to each pair formed by a tRNA T_i and an activating enzyme E_j. Discrimination in charging of tRNAs by an activating enzyme can then be defined as the ratio between the charge efficiency for the homologous tRNA and the sum of charge efficiencies for all tRNAs (with some weightings, as necessary).

Now suppose that a point variant of tRNA T_i is produced which differs from the original by a base substitution. I shall call it a first-generation variant and note it $T_i^{(1)}$. Where is this variant placed on each of the n diagrams of charging efficiency by the activating enzymes? In relation to each of the enzymes, it could have unchanged, increased or diminished efficiency. The

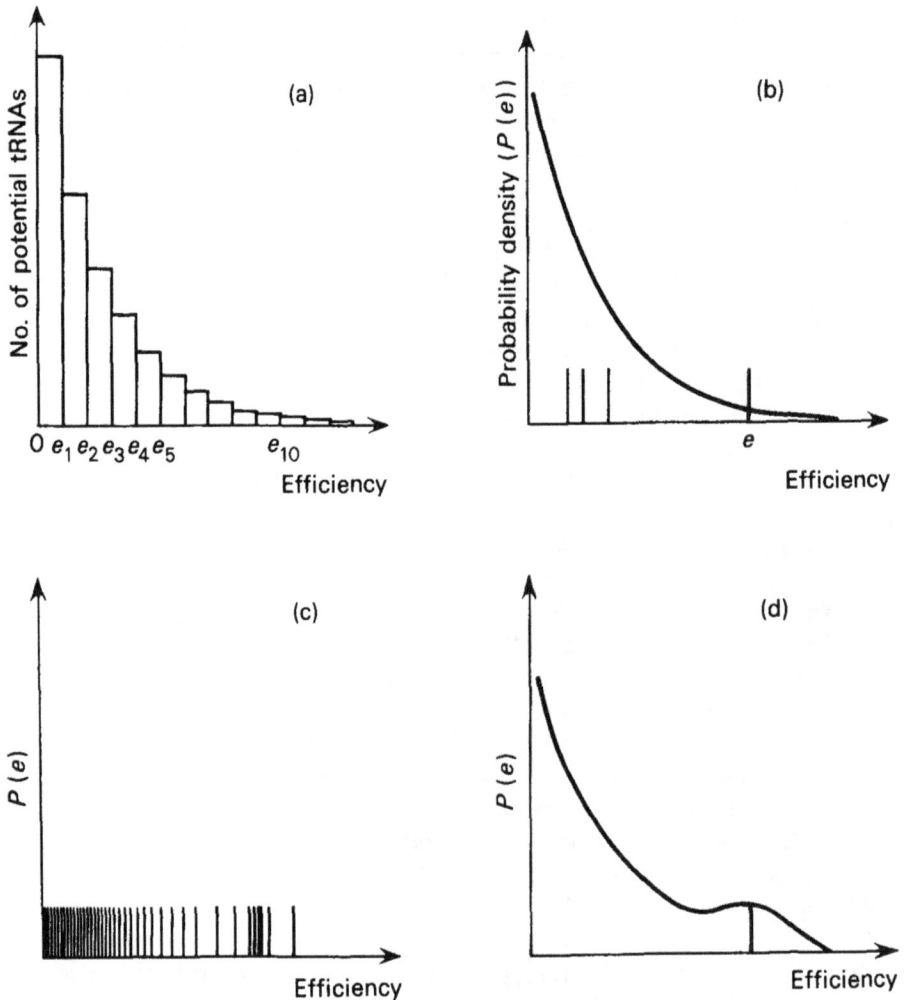

Figure 19 Real and potential tRNAs. Let us assume that we measure, for a given activating enzyme, the efficiency with which it charges each possible tRNA in tRNA space. The tRNA sequences are then distributed in classes according to the measured efficiency to give the histogram in (a). This histogram can be put into a more continuous form as in (b). The vertical lines represent the charging efficiencies of the real tRNAs of the cell by the enzyme considered. The 'specific' tRNA at the extreme right is distinguished from the non-specific tRNAs which are charged with almost negligible efficiency. If the specific tRNA mutates, it gives rise in the first generation to variants which will be charged by the homologous enzyme with varying efficiencies indicated by the narrow bars in figure (c). The density of the bars at different efficiency values gives the function F or the distribution of efficiencies for first-generation variants (d). Knowledge of this kind of function is of prime importance in dealing correctly with the stability of the code.

different possibilities can be represented by a curve which gives, for each new efficiency value, the probability density of the variants which achieve it (nevertheless, a continuous representation is no longer quite legitimate since there are only 240 first-generation variants for a tRNA eighty nucleotides long). Basically, I am introducing here the way in which the activating enzyme 'sees' the neighbourhood in tRNA space of the point corresponding to tRNA T_i and I am considering only the immediate vicinity, that of the first-generation variants. In this way I define a function F which gives the distribution of efficiencies of the enzyme E_j with the tRNAs in the immediate vicinity of T_i: the superscript (1) indicates that the efficiencies in the first generation are being considered. By definition:

$$\text{Probability that } (e_a \leqslant e(T_i^{(1)}, E_j) < e_b) = \int_{e_a}^{e_b} F_{T_i, e_j}^{(1)}(e)\, de$$

Up to now I have introduced simplifications into the calculation which did not affect the profound nature of the problem. I now make a hypothesis which is not trivial but reasonable. I assume that the curve F obtained for T_i and E_j is the same as that which would have been obtained with any other potential tRNA and the enzyme E_j, provided that the initial efficiency of the pair was the same as that of E_j with T_i. Altogether if we join all tRNAs which have the same efficiency with the enzyme E_j with a contour line in tRNA space, we can say that the enzyme sees the vicinity of each point on the contour line in the same way. Far from privileging some archetype, I am postulating invariances or equivalences which allow one to move away from the starting point without there being the slightest constraint on destination. We can now make a simplifying hypothesis which is not crucial: that the curves F which depend on the starting efficiency do not depend on the enzyme to which they refer. In other words, all activating enzymes have, at a given moment in evolution, the same statistical properties towards the set of potential tRNAs.

We can then simplify the notation and designate the function F with only two indices, one to indicate the value of e about which it applies and the other (superscript) to indicate if we are considering a first-, second- or n^{th}-generation neighbourhood. Knowing $F^{(1)}(c)$ we can deduce the second-generation function $F^{(2)}(e)$. Each first-generation variant which has been accounted for in $F^{(1)}(e)$, becomes in its turn the source of new variants. In effecting first-generation variations twice in succession, we obtain second-generation variations. $F^{(2)}$ is thus obtained by combining $F^{(1)}$ type functions. We can continue and calculate $F^{(3)}$, $F^{(4)}$, ... $F^{(\infty)}$. When n increases, the curves $F^{(n)}$ tend towards the curve initially introduced of distribution of efficiencies of the enzyme towards the set of potential tRNAs – since by varying the sequences one ends up by exploring the whole of tRNA space. So, the model requires knowledge of a single family of curves, the family $F^{(1)}(e)$. If we limit ourselves to present activating enzymes, the family of curves

$F^{(1)}(e)$ is, in principle, perfectly accessible by experiment, even though its determination demands considerable effort. The results of Giégé, Grosjean and their collaborators on interactions between the activating enzymes and tRNAs of different species indicate how efficiencies change as sequences vary. Even if the functions $F^{(1)}(e)$ are not determined experimentally, one can make reasonable estimates of them. Moreover, theoretical arguments of internal consistency could reveal constraints linking these functions.

What has been done for activating enzymes and tRNA space can be transposed to the reverse case: tRNAs and activating-enzyme space. One would then find a function $G^{(1)}(e)$ which would describe how the efficiency of the enzyme for a tRNA varies when one changes the enzyme sequence by one unit. One can continue along these lines and treat every link in the chain of information transfer in the same way. Can the existence of steady states be foreseen qualitatively? A usable principle, which takes the opposite view from Hoffmann's treatment, is that if the variant of an activating enzyme has diminished activity towards the homologous tRNA it also has lowered efficiency towards the other tRNAs and overall a better discriminating power. But I do not think that this principle will turn out to be adequate. In my opinion, a realistic model must deal with populations. The treatment which I have outlined is, in its intimate logic, stochastic. A computer simulation would probably reveal an average tendency towards disorder (which if going too far is sanctioned by the death of the individual) and from time to time gains in discriminatory power, some of which could be stabilized. A scrupulous analysis of the model would no doubt reveal links between the different functions F, G, etc. To accomplish this work would require great serenity and many years of gratuitous reflection. Then perhaps some of the more subtle links which may exist between different cellular processes would emerge clearly. But the scientist of today, caught in a system of pressures which oblige him to produce results, preferably uninteresting ones, day by day, is hardly encouraged to launch into ambitious theoretical projects.

Nevertheless, this distant objective that those studying molecular evolution are grasping for is not beyond our reach. Problems which are reputedly the most difficult mature slowly. Knowledge progresses on many fronts until the most mysterious question is answered simply by adding the last piece to a jig-saw puzzle which meanwhile had gradually been taking shape. And if you have the patience to go back over the chapters of this book and compare the contents with a work of 8 years before – *Chance and Necessity* – you will gain some measure of the speed with which not only knowledge but also fundamental concepts can evolve.

Bibliography

Bibliographic landmarks

Chapter 1: Haldane (38) Woese (55) Yčas (56).

Chapter 2: Bioenergetics (25) First organic syntheses (64) Prebiotic chemistry (18, 28, 42, 48) Original monographs of Oparin and Haldane (27) Reasons for the occurrence of the twenty amino acids (266) Molecules in space (180) Evolution of metabolic chains (147, 173, 28, 25, 88) Initiation to molecular biology (16, 23, 36) Protein structure (4, 19) Non-ribosomal syntheses of peptides (160) Steinman (42) Lability of peptides (191) Complementarity (212) Nucleic-acid structure (37) Base pairings (37, 200) The double helix (23, 50) Translation (23, 31) Regulation (16, 36) Exceptions to the code (175, 62) Selenocysteine (247) Split genes (2).

Chapter 3: The Atlas (30, 34) Comparisons of haemoglobins (288, 131, 34) in space (4, 19) Myoglobin (229) Gene sequences (35) Haemoglobin genes (105) Cytochromes (30, 113) Ultrametric property (41, 243) Haptoglobin (73) Activating enzymes (161) Highly repetitive proteins (277, 181) Abnormal haemoglobins (44) Directional effects (121) Similarity index (132) tRNA evolution (145, 164, review, 84) Mitochondrial tRNAs (97) 5S RNA (233) Archaebacteria (24) Hormones (57) Lipids (211).

Chapter 4: Reviews on protein architecture (151, 224) β–α–α (223) α_4 domain (60) Haemoglobin folding (176) Classification (172) Nucleotide fold (81) Activating enzymes (281) Immunoglobulin fold (81) Superoxide dismutase (225) Lysozyme-lactalbumin (79) Endothiapepsin (251) Pseudo-summetric proteins (182) TIDL (259, 179) Other structures with D and L amino acids (141, 77) Introns-exons (125) Fluctuations in protein structure (268) Conformational shift (263).

Chapter 5: Haemoglobin trees (30, 131) Vogel (260) Beyer (70) see also (111) Fitch (112) Principle of Parsimony (41, 53, 112, 189, 243) Stability of trees (214).

Chapter 6: Replication (43) Mutagenesis (33) Accuracy of replication (7, 67, 177) Enzymes acting on the topology of nucleic acids (122, 123) Molecular genetics (23, 36) Bernstein (68) Methylation (71) Mismatch repair (264, 127) Paramutations (110) Overlapping genes in (137) Multiple reading frames (231) Interrupted genes (2, 219) Tinkering (8) Genetic witchcraft (15, 51) Qβ (245, 209) Sumper (quoted in 5) Palindromes in ribosomal DNA (261) in viroids (232). *See also* (117).

Chapter 7: Wallace (1) Classical texts of population genetics, including Galton, Pearson and Chctverikov (40) Population genetics textbooks (20, 39, 45) The Darwinian revolution (9, 11, 13) Asymmetry (155, 256) Amplification of fluctuations (115, 77) Evolution and game theory (14, 22, 140, 174) Petit (20) Ayala (61, 66) Van Valen (258) Statistical studies on human populations (6, 10) Haldane (38) Lewontin (11) Heterozygocy levels (218) Neutralists (157, 158) Discussions (52).

Chapter 8: Naylor (190) Sulston (249, 250) Shabarova (239) Usher (257) Shamovski (217, 183) U–G (202) Ligase (188) Adenine–ribose linkage (116) Cairns-Smith (29) Pitha (216) Marlière (unpublished) Takemotot (163) Lohrmann (168) Miracle compound (150) Review (210) *See also* (255).

Chapter 9: Reference books (55, 56) Dounce (100) Regularities in the code (55, 274, 240, 186) Eck (103) Woese (271) Sonneborn (244) Crick (94) Grosjean (134) Holley (144) Crick (93) Nishimura (203) Watson (23) Ohashi (204) Reviews on translation (135, 166) Missing triplet (192) Probabilistic description (194, 195) Spirin (246) Kinetic amplifications (146, 196–8, 74) Cost of amino acids (25) Reverse translation (91) Read-through (248, 126) Tryptophan nonsense suppressor (80) Codon usage (133, 120).

Chapter 10: Interactions between amino acids and nucleic acids (82, 119, 213, 72, 171, 220) amino acids and loops (101, 55, 184, 252) polypeptides and nucleic acids (167, 170, 58, 83, 136, 237, 222, 186) Experiments (104, 234, 221) More references and reviews in (142) Epistemological argument (17) Predestination (42).

Chapter 11: Monod (16) The best of possible codes (244, 89) Crick (94) Metabolic revolution (31, 275) Woese (55) Orgel (208) Claverie and followers (87, 152, 63) Principle of Continuity (206) Take-over (29) Woese (55) Activities in statistical polypeptides (114, 99) Orgel (208) β sheets (262) Brack (76) Catalysis by synthetic polymers (156, 165, 254) Origin of the code as a repetitive process (193, 194) Three-dimensional evolution of tRNA (194) Catalytic activity of peptides (153, 215, 148, 269, 65, 138) see also (265) Eigen (106) Prigogine (21) Evolution and thermodynamics (54) Miscellaneous (272, 130, 95, 85, 274).

Chapter 12: Schuster (107) Yockey (280) Maynard Smith (242) Orthodox evolutionists (282, 49) Langride (169) Miller (185) Glass (126) *ram 1* (230) Conrad (90) Evolution through low-accuracy states (201)

Chapter 13: Primary papers (149, 69, 226, 238, 139, 276, 154) Reviews (12, 86).

Chapter 14: Ageing of amino acids (228) Proteases (109, 129) Anti-protease (241) Arms race (96) Intelligible immunology (26) Evolution of the immune response (46) Antibody evolution (205) Custom-made antibody repertoire (273) Catalytic activity in antibodies (162) Pure antibodies (168) Trypanosomes (75) Cellular evolution in higher organisms (267).

Chapter 15: Crossed-interactions of tRNAs (124) Yarus (279) Endosymbiotic theory (47, 236) Mutual adjustment of translation apparatuses (199) Primates evolution (270, 78) Hybrid haemoglobins (227, 59, 108) *See also* (278) Cook (92) Selfish genes (3) Genome organization and evolution (32) Slonimski (98).

Chapter 16: Primary papers (207, 143, 128, 187, 102) Discussions (159, 118).

References

A. Works and articles of initiation or culture

1. A. C. Brackman (1980). *A delicate arrangement. The strange case of Charles Darwin and Alfred Russel Wallace.* New York, Times Books.
2. P. Chambon (1981). Split genes. *Sci. Am.* **244,** No. 5, 60–71.
3. R. Dawkins (1976). *The selfish gene.* Oxford University Press.
4. R. E. Dickerson and I. Geis (1969). *The structure and action of proteins.* London, Harper and Row.
5. M. Eigen, W. Gardiner, P. Schuster and R. Winkler-Ostwatitsch (1981). The origin of genetic information. *Sci. Am.* **244,** No. 4, 78–94.
6. S. J. Gould (1978). Sociobiology: the art of storytelling. *N. Sci.* **80,** 530–533.
7. R. Holliday (1980). Nature's means for preserving genes. *N. Sci.* **85,** 598–600.
8. F. Jacob (1977). Evolution and tinkering. *Science* **196,** 1161–1166.
9. F. Jacob (1970). *La logique du vivant.* Paris, Gallimard.
10. A. Jacquard (1978). *Eloge de la différence: la génétique et les hommes.* Paris, Editions du Seuil.
11. R. C. Lewontin (1972). Testing the theory of natural selection. *Nature* **236,** 181–182.
12. E. C. C. Lin, A. J. Hacking and J. Aguilar (1976). Experimental models of acquisitive evolution. *BioScience* **26,** 548–555.
13. E. Mayr (1972). The nature of the darwinian revolution. *Science* **176,** 981–989.
14. E. Mayr (1974). Behaviour programs and evolutionary strategies. *Am. Sci.* **62,** 650–659.
15. A, Mendel (1980). *Les manipulations génétiques.* Paris, Editions du Seuil.
16. J. Monod (1972). *Chance and necessity: an essay on the natural philosophy of modern biology.* London, Collins.
17. J. Ninio (1976). Evolutionnisme et origines de la vie. *La Recherche* **7,** 325–334.
18. L. E. Orgel (1973). *The origins of life.* New York, Wiley.
19. M. F. Perutz (1964). The hemoglobin molecule. *Sci Am.* **211,** No. 5, 64–76.
20. C. Petit and E. Zuckerkandl (1976). *Evolution. Génétique des populations, évolution moléculaire.* Paris, Hermann.
21. I. Prigogine (1972). La thermodynamique de la view. *La Recherche* **2,** 547–562.
22. J. M. Smith (1976). Evolution and the theory of games. 2Am. Sci. **64,** 41–45.
23. J. D. Watson (1978). *Molecular biology of the gene* (3rd edn.) New York, W. A. Benjamin.
24. C. R. Woese (1981). Archaebacteria. *Sci. Am.* **244,** No. 6, 94–106.

B. Books for higher education and research

25. D. E. Atkinson (1977). *Cellular energy metabolism and its regulation.* London, Academic Press.
26. G. J. Bell, A. S. Perelson and G. H. Pimbley (Eds.) (1978). *Theoretical immunology.* New York, Basel, Marcel Dakker.
27. J. D. Barnal (1967). *The origin of life.* London, Weidenfeld and Nicolson.
28. R. Buvet (1974). *L'origine des êtres vivants et des processus biologiques.* Paris, Masson.
29. A. G. Cairns-Smith (1982). *Genetic Takeover.* Cambridge, Cambridge University Press.
30. M. O. Dayhoff (1972). *Atlas of protein sequence and structure,* vol. 5; (1973) suppl. 1; (1976) suppl. 2. Silver Spring, Maryland, The National Biochemical Research Foundation.
31. L. S. Dillon (1978). *The genetic mechanism and the origin of life.* New York, Plenum Press.
32. G. A. Dover and R. B. Flavell (Eds.) (1982). *Genome evolution.* London, Academic Press.
33. J. W. Drake (1970). *The molecular basis of mutation.* San Francisco, London, Cambridge, Amsterdam, Holden-Day.
34. R. Eck and M. O. Dayhoff (1969). *Atlas of protein sequence and structure,* vol. 4. Silver Spring, Maryland, The National Biomedical Research Foundation.
35. C. Gautier, M. Gouy, M. Jacobzone and R. Grantham (1981). *Nucleic acid sequences handbook.* New York, Praeger.
36. R. E. Glass (1982). *Gene function. E.* coli *and its inheritable elements.* London, Croom Helm.
37. W. Guschlbauer (1976). *Nucleic acid structure.* Berlin, Springer-Verlag.
38. J. B. S. Haldane (1932). *The causes of evolution.* New York, Harper. Reprinted by Cornell University Press (1966), Ithaca, New York.
39. A. Jacquard (1977). *Concepts en génétiques des populations.* Paris, Masson.
40. D. L. Jameson (Ed.) (1977). *Evolutionary genetics.* Stroudsburg (Pennsylvania), Dowden, Hutchinson and Ross.
41. N. Jardine and R. Sibson (1971). *Mathematical taxonomy.* New York, Wiley.
42. D. H. Kennyon and G. Steinman (1969). *Biochemical predestination.* London, McGraw-Hill.
43. A. Kornberg (1980). *DNA replication.* San Francisco, Freeman.
44. H. Lehman and P. A. M. Kynoch (1976). *Human haemoglobin variants and their characteristics.* Amsterdam, North-Holland.
45. R. C. Lewontin (1974). *The genetic basis of evolutionary change.* New York, Columbia University Press.
46. J. J. Marchalonis (1977). *Immunity in evolution.* London, Edward Arnold.
47. L. Margulis (1981). *Symbiosis in cell evolution. Life and its environment on the early earth.* San Francisco, Freeman.
48. S. Miller and L. E. Orgel (1974). *The origins of life on the earth.* London, Prentice-Hall.
49. S. Ohno (1970). *Evolution by gene duplication.* Berlin, Springer-Verlag.
50. R. Olby (1974). *The path to the double helix.* London, MacMillan.
51. G. de Pawlowski (1973). *Fallait y penser.* Paris, Balland.

52. J. M. Smith and R. Holliday (Eds.) (1979). *The evolution of adaptation by natural selection. Proceedings of the Royal Society of London, Series B*, Vol. 205. London. (Also published as an autonomous book by The Royal Society.)
53. R. R. Sokal and P. H. A. Sneath (1973). *Principles of numerical taxonomy*. San Francisco, Freeman.
54. J. Tonnelat (1978 and 1979). *Thermodynamique et biologie*, vol. 1: (*Entropie, désordre et complexité*), vol. 2: (*l'ordre issu du hasard*). Paris, Maloine.
55. C. R. Woese (1967). *The genetic code*. New York, Evanston, London, Harper.
56. M. Yčas (1969). *The biological code*. Amsterdam, London, North-Holland.

C. Original papers for specialists

57. R. Acher (1980). *Proc. Roy Soc. B* 210, 21–43.
58. K. Adler, *et al.* (1972). *Nature* **237**, 322–327.
59. E. Antonini, *et al.* (1965). *Biochim. Biophys. Acta* **104**, 160–166.
60. P. Argos, M. G. Rossmann and J. E. Johnson (1977). *Biochem. Biophys. Res. Commun.* **75**, 83–86.
61. F. J. Ayala (1969). *Proc. Natl. Acad. Sci. USA* **63**, 790–793.
62. B. G. Barrell, A. T. Bankier and J. Drouin (1979). *Nature* **282**, 189–194.
63. N. A. Barricelli (1977). *J. Theoret. Biol.* **67**, 85–109.
64. H. Bauer (1980). *Naturwissenschaften* **67**, 1–6.
65. T. Behmoaras, J.-J. Toulmé and C. Hélène (1981). *Nature* **292**, 858–859.
66. M. B. Benado, F. J. Ayala and M. M. Green (1976). *Genetics* **82**, 43–52.
67. F. Bernardi and J. Ninio (1978). *Biochimie* **60**, 1083–1095.
68. H. Bernstein (1977). *J. Theoret. Biol.* **69**, 371–380.
69. J. L. Betz, P. R. Brown, M. J. Smyth and P. H. Clarke (1974). *Nature* **247**, 261–264.
70. W. A. Beyer, M. L. Stein, T. F. Smith and S. W. Ulam (1974). *Math. Biosci.* **19**, 9–25.
71. A. P. Bird (1978). *J. Mol. Biol.* **119**, 49–60.
72. S. Black (1973). *Adv. Enzymol.* **38**, 193–234.
73. J. A. Black and G. H. Dixon (1970). *Can. J. Biochem.* **48**, 133–146.
74. C. Blomberg (1977). *J. Theoret. Biol.* **66**, 307–325.
75. P. Borst and M. Cross (1982). *Cell* **29**, 291–303.
76. A. Brack and L. E. Orgel (1975). *Nature* **256**, 383–387.
77. A. Brack and G. Spach (1980). *J. Mol. Evol.* **15**, 231–238.
78. W. M. Brown, E. M. Prager, A. Wang and A. C. Wilson (1982). *J. Mol. Evol.* **18**, 225–239.
79. W. J. Browne, *et. al.* (1969). *J. Mol. Biol.* **42**, 65–86.
80. R. H. Buckingham and C. G. Kurland (1977). *Proc. Natl. Acad. Sci. USA* **74**, 5496–5498.
81. M. Buchner, G. E. Ford, D. Moras, K. W. Olsen and M. G. Rossman (1974). *J. Mol. Biol.* **90**, 25–49.
82. P. C. Caldwell and C. Hinshelwood (1950). *J. Chem. Soc.* **4**, 3156–3159.
83. C. W. Carter and J. Kraut (1974). *Proc. Natl. Acad. Sci. USA* **71**, 283–287.
84. R. J. Cedergren, D. Sankoff, B. La Rue and H. Grosjean (1981). *C.R.C. Crit. Rev. in Biochem.* **11**, 35–104.

85. D. S. Chernavskii and N. M. Chernavskaya (1975). *J. Theoret. Biol.* **50**, 13–23.
86. P. Clarke (1978). In Gunsalus, I. C., Ornaston, L. N., and Sokatch, J. R. (Eds.): *The bacteria. A treatise on structure and function*, vol. 6: *bacterial diversity*, pp. 137–218. London, Academic Press.
87. P. Claverie (1971). *J. Mol. Biol.* **56**, 75–82.
88. G. N. Cohen, I. Saint-Girons and P. Truffa-Bachi (1977). *Trends Biochem. Sci.* **2**, 97–99.
89. M. Conrad (1970). *Currents in Modern Biol.* **3**, 260–264.
90. M. Conrad (1982). *BioSystems* **15**, 83–85.
91. N. D. Cook (1977). *J. Theoret. Biol.* **64**, 113–135.
92. R. A. Cook and D. E. Koshland, Jr. (1969). *Proc. Natl. Acad. Sci. USA* **64**, 247–254.
93. F. H. C. Crick (1966). *J. Mol. Biol.* **19**, 548–555.
94. F. H. C. Crick (1968). *J. Mol. Biol.* **38**, 367–379.
95. F. H. C. Crick, S. Brenner, A. Klug and G. Piecznik (1976). *Origins of Life* **7**, 389–397.
96. R. Dawkins and J. R. Krebs (1979). *In* Smith, J. M. and Holliday, R. (Eds.): *The evolution of adaptation by natural selection. Proceedings of the Royal Society of London, Series B*, vol. 205, pp. 489–511. London.
97. M. H. L. de Bruijn, P. H. Schreier, I. C. Eperton and B. G. Barrell (1980). *Nucleic Acids Res.* **8**, 5213–5222.
98. H. de la Salle, C. Jacq and P. P. Slonimski (1982). *Cell* **28**, 721–732.
99. K. Dose (1976). *In Protein Structure and Evolution*, pp. 149–184. New York, Basel, Marcel Dekker.
100. A. L. Dounce (1952). *Enzymologia* **15**, 251–258.
101. P. Dunnill (1966). *Nature* **210**, 1267–1268.
102. F. J. Dyson (1982). *J. Mol. Evol.* **18**, 344–350.
103. R. Eck (1963). *Science* **140**, 477–481.
104. H. Eckstein, H. Schott and E. Bayer (1976). *Biochim. Biophys. Acta* **432**, 1–9.
105. A. Efstratiadis *et al.* (1980). *Cell* **21**, 653–668.
106. M. Eigen (1971). *Naturwissenschaften* **58**, 465–523.
107. M. Eigen and P. Schuster (1977). *Naturwissenschaften* **64**, 541–565.
108. Y. Enoki and S. Tomita (1968). *J. Mol. Biol.* **32**, 121–134.
109. S. Erhan and L. D. Greller (1974). *Nature* **251**, 353–354.
110. M. S. Esposito and R. E. Esposito (1977). *In:* Goldstein, L. and Prescott, D. M. (Eds.): *Cell biology: a comprehensive treatise*, vol. 1: *genetic mechanisms of cells*, pp. 59–92. London, Academic Press.
111. J. S. Farris (1972). *Am. Nat.* **106**, 645–668.
112. W. M. Fitch (1977). *Am. Nat.* **111**, 223–257.
113. W. M. Fitch and E. Margoliash (1967). *Science* **155**, 279–284.
114. F. Fox (1976). *In: Protein structure and evolution*, pp. 125–148. New York, Basel, Marcel Dekker.
115. F. C. Frank (1953). *Biochim. Biophys. Acta* **11**, 459–469.
116. W. D. Fuller, R. A. Sanchez and L. E. Orgel (1972). *J. Mol. Biol.* **67**, 24–33.
117. D. J. Galas (1978). *J. Theoret. Biol.* **72**, 57–73.
118. J. A. Gallant and J. Prothero (1980). *J. Theoret. Biol.* **83**, 561–578.
119. Gamow (1954). *Nature* **173**, 318.
120. J.-P. Garel (1982). *Trends in Biochem. Sci.* **7**, 105–108.

121. C. Gautier (1976). *Science* **194**, 642.

122. K. Geider and H. Hoffmann-Berling (1981). *Ann. Rev. Biochem.* **50**, 233–260.

123. M. Gellert (1981). *Ann. Rev. Biochem.* **50**, 879–910.

124. R. Grégé, D. Kern, J.-P. Ebel, H. Grosjean, S. de Hénau and H. Chantrenne (1974). *Eur. J. Biochem.* **45**, 351–362.

125. W. Gilbert (1978). *Nature* **271**, 501 only.

126. R. E. Glass, V. Nene and M. G. Hunter (1982). *Biochem. J.* **203**, 1–13.

127. B. W. Glickman and M. Radman (1980). *Proc. Natl. Acad. Sci. USA* **77**, 1063–1067.

128. N. S. Goel and S. Islam (1977). *J. Theoret. Biol.* **57**, 167–182.

129. A. L. Goldberg (1976). *Ann. Rev. Biochem.* **45**, 747–803.

130. A. L. Goldberg and R. E. Wittes (1966). *Science* **153**, 420–424.

131. M. Goodman, W. Moore and G. Matsuda (1975). *Nature* **253**, 603–608.

132. R. Grantham (1974). *Science* **185**, 862–864.

133. R. Grantham, C. Gautier and M. Gouy (1980). *Nucleic Acids Res.* **8**, 1893–1912.

134. H. Grosjean, S. de Hénau and D. Crothers (1978). *Proc. Natl. Acad. Sci. USA* **75**, 610–614.

135. M. Grunberg-Manago, R. H. Buckingham, B. S. Cooperman and J. W. B. Hershey (1978). In Stanier, R. Y., Rogers, H. J. and Ward, J. B. (Eds.): *Relations between structure and function in the prokaryotic cell*, pp. 27–110. Cambridge University Press.

136. G. V. Gursky, *et al.* (1976). *Mol. Biol. Rep.* **2**, 413–425.

137. G. N. Gussin, K.-M. Yen and L. F. Reichardt (1975). *Virology* **63**, 273–277.

138. B. Gutte, M. Daümigen and E. Wittschieber (1979). *Nature* **281**, 650–655.

139. B. G. Hall (1981). *Biochemistry* **20**, 4042–4049.

140. W. D. Hamilton (1964). *J. Theoret. Biol.* **7**, 1–16, 17–52.

141. F. Heitz, B. Lotz and G. Spach (1975), *J. Mol. Biol.* **92**, 1–13.

142. C. Hélène and G. Lancelot (1982). *Progress in Biophysics and Mol. Biol.* **39**, 1–68.

143. G. W. Hoffman (1974). *J. Mol. Biol.* **86**, 349–362.

144. R. W. Holley (1966). *Sci. Am.* **214**, 30–38.

145. R. Holmquist, T. H. Jukes and S. Pangburn (1973). *J. Mol. Biol.* **78**, 91–116.

146. J. J. Hopfield (1974). *Proc. Natl. Acad. Sci. USA* **71**, 4135–4139.

147. N. H. Horowitz (1945). *Proc. Natl. Acad. Sci. USA* **31**, 153–157.

148. Y. Imanishi (1978). *Adv. Polymer Sci.* **20**, 1–77.

149. C. B. Inderlied and R. P. Mortlock (1977). *J. Mol. Evol.* **9**, 181–190.

150. T. Inoue and L. E. Orgel (1982). *J. Mol. Biol.*, in press.

151. J. Janin (1977). *Bull. Inst. Pasteur* **77**, 337–373.

152. T. H. Jukes (1973). *Nature* **246**, 22–26.

153. A. Kapoor (1972). *In* Lande, S. (Ed.): *Progress in peptide research*, vol. 2, pp. 335–341. New York, Gordon and Breach.

154. J. Kemper (1974). *J. Bact.* **120**, 1176–1185.

155. L. Keszthelyi (1977). *Origins of Life* **8**, 299–340.

156. H. C. Kieffer, W. I. Congdon, I. S. Scarpa and I. Klotz (1972). *Proc. Natl. Acad. Sci. USA* **69**, 2155–2159.

157. M. Kimura (1968). *Nature* **217**, 624–626.

158. J. L. King and T. H. Jukes (1969). *Science* **164**, 788–798.

159. T. B. L. Kirkwood (1980). *J. Theoret. Biol.* **82**, 363–382.

160. H. Kleinkauf and H. Koischwitz (1978). *In: Progress in molecular and subcellular biology*, vol. 6, pp. 59–112. Berlin, Springer-Verlag.

161. G. L. E. Koch, Y. Boulanger and B. S. Hartley (1974). *Nature* **249**, 316–320.

162. F. Kohen, J. B. Kim, H. R. Lindner, Z. Eshaar and B. Green (1980). *FEBS Letts.* **111**, 427–431.

163. K. Kondo, S. Tanioku and K. Takemoto (1980). *Makromol. Chem., Rapid Commun.* **1**, 303–306.

164. E. Kubli (1980). *Trends Biochem. Sci.* **5**, 90–91.

165. T. Kunitake and Y. Okahata (1976). *Adv. Polymer Sci.* **24**, 1–87.

166. C. G. Kurland (1980). *In Chambliss, G. et al.* (Eds.): *Ribosomes. Structure, function and genetics*, pp. 597–614. Baltimore, University Park Press.

167. J. C. Lacey and K. M. Pruitt (1969). *Nature* **223**, 799–804

168. D. Lane and H. Koprowski (1982). *Nature* **296**, 200–201.

169. J. Langridge (1968). *J. Bacteriol.* **96**, 1711–1717.

170. J. Lejeune (1977). *C.R. Acad. Sci., série D.* **285**, 249–252.

171. A. Lesk (1970). *J. Theoret. Biol.* **27**, 171–173.

172. M. Levitt and C. Chothia (1976). *Natur* **261**, 552–557.

173. E. B. Lewis (1951). *Cold Spring Harbor Symp. Quant. Biol.* **16**, 159–174.

174. R. C. Lewontin (1961). *J. Theoret. Biol.* **1**, 382–403.

175. M. Li and A. Tzagoloff (1979). *Cell* **18**, 47–53.

176. V. I. Lim and A. V. Efimov (1977). *FEBS Letts* **78**, 279–283.

177. L. A. Loeb and T. A. Kunkel (1982). *Ann. Rev. Biochem.* **51**, 429–457.

178. L. R. Lohrmann and L. E. Orgel (1976). *Natur* **261**, 342–344.

179. B. Lotz, F. Colonna-Cesari, F. Heitz and G. Spach (1976). *J. Mol. Biol.* **106**, 915–942.

180. A. P. C. Mann and D. A. Williams (1980). *Nature* **283**, 721–725.

181. A. D. McLachlan (1977). *Biopolymers* **16**, 1271–1297.

182. A. McLachland (1980). *In Jaenicke, R.* (Ed.): *Protein folding*, pp. 79–96. Amsterdam, Elsevier/North Holland Biomedical Press.

183. N. V. Melamed, S. G. Popov and G. G. Shamovsky (1974). *Izv. Sib. Otd. Akad. Nauk SSSR Ser Khim. Nauk* **5**, 90–98.

184. G. Melcher (1974). *J. Mol. Evol.* **3**, 121–140.

185. J. H. Miller (1980). *In J. H. Miller and W. S. Reznikoff* (Eds.): *The operon*, pp. 31–88. New York, Cold Spring Harbor Laboratory.

186. M. Mitsuyama (1982). *Bull. Inst. Pasteur* **82**, 728–742.

187. H. Mizutani and C. Ponnamperuma (1977). *Origins of Life* **8**, 183–219.

188. P. Modrich and I. R. Lehman (1973). *J. Biol. Chem.* **248**, 7502–7511.

189. G. W. Moore (1977). *J. Theoret. Biol.* **66**, 95–106.

190. R. Naylor and P. T. Gilham (1966). *Biochemistry* **8**, 2722–2728.

191. I. Nicholson (1970). *J. Macromol. Sci. Chem.* **A4(7)**, 1619–1625.

192. J. Ninio (1971). *J. Mol. Biol.* **56**, 63–74.

193. J. Ninio (1971). *J. Mol. Evol.*, rejected.

194. J. Ninio (1973). *Progr. Nucleic Acid Res. Mol. Biol.* **13**, 301–337.

195. J. Ninio (1974). *J. Mol. Biol.* **84**, 297–313.

196. J. Ninio (1974). *Nature*, rejected.

197. J. Ninio (1975). *In Sadon, C.* (Ed.): *Ecole de Rocsoff 1974. L'évolution des macromolécules biologiques*, pp. 51–68. Paris, C.N.R.S.

198. J. Ninio (1975). *Biochimie* **57**, 587–595.

199. J. Ninio (1975). *In* Puiseux-Dao (Ed.): *Molecular biology of nucleocytoplasmic relationship*, pp. 31–39. Amsterdam, Elsevier.
200. J. Ninio (1979). *Biochemie* **61**, 1133–1150.
201. J. Ninio (1981). *In* Walcher, D. and Kretchmer, N. (Eds.): *Food, nutrition and evolution*, pp. 9–13. New York, Masson Publishing USA.
202. J. Ninio and L. E. Orgel (1978). *J. Mol. Evol.* **12**, 91–99.
203. S. Nishimura (1972). *Progr. in Nucleic Acid Res. Mol. Biol.* **12**, 49–85.
204. Z. Ohashi, N. Saneyoshi, F. Harada, H. Harada and S. Nishimura (1970). *Biochem-Biophys. Res. Commun.* **40**, 866–872.
205. T. Ohta (1978). *Proc. Natl. Acad. Sci. USA* **75**, 5108–5112.
206. L. E. Orgel (1968). *J. Mol. Biol.* **38**, 281–393.
207. L. E. Orgel (1970). *Proc. Natl. Acad. Sci. USA* **67**, 1476.
208. L. E. Orgel (1972). *Isr. J. Chem.* **10**, 287–292.
209. L. E. Orgel (1979). *In* Smith, J. M. and Holliday, R. (Eds.): *The evolution of adaptation by natural selection. Proceedings of the Royal Society of London. Series B*, **205**, pp. 435–442. London.
210. L. E. Orgel and R. Lohrmann (1974). *Accounts of Chem. Res.* **7**, 368–377.
211. G. Ourisson, P. Albrecht and M. Rohmer (1982). *Trends in Biochem. Sci.* **7**, 236–239.
212. L. Pauling and M. Delbrück (1940). *Science* **92**, 77–79.
213. S. R. Pelc and M. G. E. Welton (1966). *Nature* **209**, 868–870.
214. D. Penny, L. R. Foulds and M. D. Hendy (1982). *Nature* **277**, 197–200.
215. D. Petz and F. Schneider (1976). *FEBS Letts.* **67**, 32–35.
216. P. M. Pitha and J. Pitha (1972). *Nature New Biol.* **240**, 78–80.
217. S. G. Popov, G. G. Shamovsky, S. I. Eremenko and J. M. Backer (1970). *Biophyzika* **21**, 739–741.
218. J. A. M. Ramshaw, J. A. Coyne and R. C. Lewontin (1979). *Genetics* **93**, 1019–1037.
219. D. Reanney (1979). *Nature* **227**, 598–600.
220. R. Rein, R. Garduno, J. T. Egan, S. Columbano (1977). *BioSystems* **9**, 131–137.
221. J. Reuben (1978). *FEBS Letts.* **94**, 20–24.
222. A. Rich (1962). *Acta Pontifica, Vatican* **22**, 271–284.
223. J. S. Richardson (1976). *Proc. Natl. Acad. Sci. USA* **73**, 2619–2623.
224. J. S. Richardson (1981), *Adv. in Prot. Chem.* **34**, 167–339.
225. J. S. Richardson, D. C. Richardson, K. A. Thomas, E. W. Silverton and D. R. Davies (1976). *J. Mol. Biol.* **102**, 221–235.
226. P. W. J. Rigby, B. D. Burley, Jr. and B. S. Hartley (1974). *Nature* **251**, 200–204.
227. A. Riggs and A. E. Herner (1962). *Proc. Natl. Acad. Sci. USA* **48**, 1664–1669.
228. A. B. Robinson (1974). *Proc. Natl. Acad. Sci. USA* **71**, 885–888.
229. A. E. Romero-Herrara, H. Lehman, K. A. Joysey and A. E. Friday (1978). *Phil. Trans. R. Soc. Series B* **283**, 61–163.
230. R. Rosset and L. Gorini (1969). *J. Mol. Biol.* **39**, 95–112.
231. F. Sanger *et al.* (1977). *Nature* **265**, 687–695.
232. H. L. Sanger *et al.* (1976). *Proc. Natl. Acad. Sci. USA* **72**, 3852–3856.
233. D. Sankoff, C. Morel and R. J. Cedergren (1973). *Nature New Biol.* **245**, 232–234.
234. C. Saxinger and C. Ponnamperuma (1971). *J. Mol. Evol.* **1**, 63–73.

235. M. Schiffer, R. L. Girling, K. R. Ely and A. B. Edmundson (1973). *Biochemistry* **12**, 4620–4631.
236. R. M. Schwartz and M. O. Dayhoff (1978). *Science* **199**, 395–403.
237. N. C. Seeman, J. M. Rosenberg and A. Rich (1976). *Proc. Natl. Acad. Sci. USA* **73**, 804–808.
238. E. Senor, A. T. Bull and J. H. Slater (1976). *Nature* **263**, 476–479.
239. Z. Shabarova and M. Prokofiev (1970). *FEBS Letts.* **11**, 237–240.
240. I. Z. Siemion and A. Paradowski (1980). *In* J. Augustinyak (Ed.): *Biological implications of protein-nucleic acid interactions*, p. 90. Amsterdam, Elsevier/North-Holland Biomedical Press.
241. L. D. Simon, K. Tomcozak and A. C. St. John (1978). *Nature* **275**, 424–428.
242. J. M. Smith (1970). *Nature* **225**, 563–564.
243. P. H. A. Sneath (1974). *In* Carlile, M. J. and Skehel, J. J. (Eds.): *Evolution in the microbial world*, pp. 1–39. Cambridge University Press.
244. T. M. Sonneborn (1965). *In* Bryson, V. and Vogel, H. (Eds.): *Evolving genes and proteins*, pp. 377–397. London, Academic Press.
245. S. Spiegelman (1971). *Q. Rev. Biophys.* **4**, 213–253.
246. A. S. Spirin (1978). *Progr. in Nucleic Acid Res. Mol. Biol.* **21**, 39–62.
247. T. C. Stadtman (1980). *Trends Biochem. Sci.* **5**, 203–206.
248. D. A. Steege and D. G. Söll (1979). *In* Goldberger, R. F. (Ed.): *Biological regulation and development*, vol 1: *Gene expression*, pp. 433–485. New York, Plenum Press.
249. J. Sulston, R. Lohrmann, L. E. Orgel and T. H. Miles (1968). *Proc. Natl. Acad. Sci. USA* **59**, 726–733.
250. J. Sulston *et al.* (1969). *J. Mol. Biol.* **40**, 227–234.
251. J. Tang, M. N. G. James, I. N. Hsu, J. A. Jenkins and T. L. Blundell (1978). *Nature* **271**, 618–621.
252. B. R. Thomas (1970). *Biochem. Biophys. Res. Commun.* **40**, 1289–1296.
253. M. D. Topal and J. R. Fresco (1976). *Nature* **263**, 285–289.
254. E. Tsuchida and H. Nishide (1977). *Adv. Polymer Sci.* **24**, 1–87.
255. S. Uesugi and M. Ikehara (1977). *Biochemistry* **16**, 493–498.
256. T. L. V. Ulbricht (1981). *Symbioses* **13**, 60–71.
257. D. A. Usher and A. H. McHale (1976). *Proc. Natl. Acad. Sci. USA* **73**, 1149–1153.
258. L. Van Valen (1974). *J. Mol. Evol.* **3**, 89–101.
259. W. R. Veatch, E. T. Fossell and E. R. Blout (1974). *Biochemistry* **13**, 5249–5256.
260. H. Vogel (1975). *In* Sadron, C. (Ed.): *Ecole de Roscoff 1974. L'évolution des macromolécules biologiques*, pp. 81–93. Paris, C.N.R.S.
261. V. M. Vogt and R. Braun (1976). *J. Mol. Biol.* **106**, 567–587.
262. G. von Heijne and C. Blomberg (1978). *Biopolymers* **17**, 2033–2037.
263. G. von Heijne, C. Blomberg and H. Baltscheffsky (1977). *Origins of Life* **9**, 27–37.
264. R. Wagner, Jr. and M. Meselson (1976). *Proc. Natl. Acad. Sci. USA* **73**, 4135–4139.
265. J. A. Walder, R. Y. Walder, M. J. Heller, S. M. Freier, R. L. Letsinger and I. M. Klotz (1979). *Proc. Natl. Acad. Sci. USA* **76**, 51–55.
266. A. L. Weber and S. L. Miller (1981). *J. Mol. Evol.* **17**, 273–284.
267. J.-C. Weill and C.-A. Reynaud (1980). *BioSystems* **12**, 23–25.

268. G. R. Welch, B. Somogyi and S. Damjanovich (1982). *Progress in Biophys. Mol. Biol.* **39,** 109–146.
269. D. Werner, S. A. Russel and H. J. Evans (1973). *Proc. Natl. Acad. Sci. USA* **70,** 339–342.
270. A. C. Wilson, S. S. Carlson and T. J. White (1977). *Ann. Rev. Biochem.* **46,** 573–639.
271. C. R. Woese (1962). *Nature* **194,** 1114–1115.
272. C. R. Woese (1973). *Naturwissenschaften* **60,** 447–459.
273. C. R. Woese (1974). *J. Mol. Evol.* **3,** 109–113.
274. R. V. Wolfenden, P. M. Cullis and C. C. F. Southgate (1979). *Science* **206,** 575–577.
275. J. T. Wong (1975). *Proc. Natl. Acad. Sci. USA* **72,** 1909–1912.
276. T. T. Wu (1976). *Biochim. Biophys. Acta* **128,** 656–663.
277. M. Yčas (1976). *Fed. Proc.* **35,** 2139–2140.
278. C. Yanofsky, S. S.-L. Li, V. Horn and J. Rowe (1977). *Proc. Natl. Acad. Sci. USA* **74,** 286–290.
279. M. Yarus, R. Knowlton and L. Soll (1977). *In* Vogel, H. J. (Ed.): *Nucleic acid-protein recognition*, pp. 391–408. London, Acadamic Press.
280. H. P. Yockey (1977). *J. Theoret. Biol.* **67,** 377–398.
281. C. Zelwer, J.-L. Risler and S. Brunie (1982). *J. Mol. Biol.* **155,** 63–81.
282. E. Zuckerkandl and L. Pauling (1965). *In* Bryson, V. and Vogel, H. J. (Eds): *Evolving genes and proteins*, pp. 97–166. London, Academic Press.

GPSR Authorized Representative: Easy Access System Europe - Mustamäe tee
50, 10621 Tallinn, Estonia, gpsr.requests@easproject.com

www.ingramcontent.com/pod-product-compliance
Lightning Source LLC
Chambersburg PA
CBHW082008190326
41458CB00010B/3116